NUMERICAL MODELLING OF ICE FLOODS IN THE NING-MENG REACH

OF THE YELLOW RIVER BASIN

NUMERICAL MODELLING OF ICE FLOODS IN THE NING-MENG REACH

OF THE YELLOW RIVER BASIN

DISSERTATION

Submitted in fulfillment of the requirements of
the Board for Doctorates of Delft University of Technology
and of
the Academic Board of the UNESCO-IHE Institute for Water Education
for the
Degree of DOCTOR
to be defended in public on
Friday 8 December, 2017 at 10:00 hours
in Delft, the Netherlands

by

Chunqing WANG

Master of Science in Hydroinformatics
UNESCO-IHE Institute for Water Education
Delft, the Netherlands
Born in Henan, China

This dissertation has been approved by the
Promotor: Prof.dr.ir. A.E. Mynett
Copromotor: Dr.ir. I. Popescu

Composition of Doctoral Committee:

Chairman	Rector Magnificus, Delft University of Technology
Vice-Chairman	Rector UNESCO-IHE
Prof.dr.ir. A.E. Mynett	UNESCO-IHE / Delft University of Technology, promotor
Dr.ir. I. Popescu	UNESCO-IHE, copromotor

Independent members:

Prof.dr.ir. S.N. Jonkman	Delft University of Technology
Prof.dr.ir. J.A. Roelvink	UNESCO-IHE / Delft University of Technology
Prof.dr. R.A. Falconer	Cardiff University, UK
Prof.dr. W. Wang	Hohai University, China
Prof.dr.ir. W.S.J. Uijttewaal	Delft University of Technology, reserve member

CRC Press/Balkema is an imprint of the Taylor & Francis Group, an informa business

Published by:
CRC Press/Balkema
Schipholweg 107C, 2316 XC, Leiden, the Netherlands
Pub.NL@taylorandfrancis.com
www.crcpress.com – www.taylorandfrancis.com
ISBN 978-1-138-48701-7 (Taylor & Francis Group)

This research was supported by the
Yellow River Conservancy Commission

Summary

Summary (English)

Ice is present during part of the year on many rivers in cold and even temperate regions of the globe. Although largely ignored in hydrological literature, river ice can have serious impacts, including extreme flood events triggered by ice jams, interference with transportation and energy production, reduced river flows and associated ecological and water quality consequences. Ice is a significant factor influencing planetary biogeochemical cycles and the development of certain ecosystems. River ice phenomena include the formation, evolution, transport, accumulation, dissipation, and deterioration of various forms of ice. River ice processes involve complex interactions between hydrodynamic, mechanical and thermal processes, which are also influenced by meteorological and hydrological conditions. The occurrence of ice in rivers is an important phenomenon to be considered in the development of water resources in cold regions. Ice formation can affect the design, operation and maintenance of reservoirs. Major engineering concerns related to river ice are ice jamming, reservoir operations, water transfer, and environmental and morphological effects.

The Ning-Meng reach (including Ningxia Hui Nationality Autonomous Region and Inner Mongolia Autonomous Region) is located at the Northern part of the Yellow River basin and has a length of 1,237 km. Due to its special geographical location and river flow direction, the Ning-Meng reach freezes up every year in winter. Both during the freeze-up and breakup period, unfavourable conditions can occur which may cause ice jams and ice dams to occur, leading to dike breaching and overtopping of the embankment, which has resulted in huge casualties and property losses throughout history. Following the development of the integrated water resources management plan for the Yellow River, the requirements for water regulation in the upper Yellow River during ice flood periods should not only safeguard against ice floods, but also assure the availability of limited water resources. This determines the overall requirement for ice regime forecasting including lead-time and precision. In order to solve the above mentioned problem, a numerical model is one of the essential parts of the current research going on at the Yellow River Conservancy Commission (YRCC), which can be used to supplement the inadequacies in the field and lab studies which are being carried out to help understand the physical processes related to river ice on the Yellow River.

Although numerical ice flood models have been built for several rivers in the world, they mainly have a simulation purpose and are often applied only over short distances. Also, they are not designed to make forecasts and usually lack the capability to deal with complex flow patterns and river geometries. Due to the complexity of various river ice phenomena in each period, it is necessary to calculate both the water balance and thermal balance resp. according to different criteria and to adjust the water balance equations for the effects of ice phenomena (e.g. in the continuity equation). Enhanced economic development and human activities have changed the characteristics of ice regimes in recent decades, especially ice disasters during the freezing or breaking-up periods. Hence, it is very important to know ice regime characteristics and use mathematical models to enable ice flood forecasting, ice flood prevention and ice flood alleviation.

In this research, hydrological and meteorological data from 1950 to 2010 have been used to analyse the characteristics of ice regimes, especially after the Liujiaxia reservoir was put into operation in 1986. Some data were used for river ice modelling, in particular coefficient and parameter verification which are critical for solving key problems during river ice model setup. Furthermore, additional field observations were carried out for ice flood model calibration and validation. By combining meteorological forecast models and statistical forecast models to setup a medium and short range air temperature forecast model of the Ning-Meng reach, the results could be used for improved ice formation forecast and prolong the predictive lead-time of numerical ice flood models. Considering the winter situations of the Ning-Meng reach, added channel water storage terms are needed in the continuity equation to account for ice cover formation and floodplain water storage that affect the mass balance. For the momentum equation, an ice cover friction term should be added in the momentum balance. The proposed channel water storage terms not only maintain the mass balance, but also provide ways to calculate the channel water storage for supporting the reservoir regulation during the ice breakup period. The thickness of the ice layer on the water surface is equal to the thickness of the ice block floating on the water surface with a coefficient of submergence of 0.9. For the water temperature and ice run simulations, a simplified formula is proposed for effective calculation. An empirical criterion is proposed based on air temperature, channel geometry, discharge, and ice cover thickness, to judge whether river freeze-up or breakup may occur. Using these calibrated parameters together with a 1-10 days forecasted air temperature proved very effective to predict river freeze-up and breakup with a long lead-time.

The one-at-a-time sensitivity method was used to conduct a sensitivity analysis of the parameters for ice cover roughness, decay constant, heat exchange coefficient between water and atmosphere, and water temperature. The river freeze-up and breakup criteria were calibrated using the observed hydro-meteorological data. Based on this, the uncertainty analysis distinguished between uncertainty about water level and ice cover thickness at Sanhuhekou station. A Monte Carlo simulation method was used to conduct a parametric uncertainty analysis.

The overall finding was that the numerical ice flood model developed in this thesis for the Ning-Meng reach can be applied to simulate water temperature, ice run concentrations, ice cover thickness, river freeze-up and breakup time, flow discharge, water level and channel water storage. The ice flood model simulation results proved to be acceptable and reasonable, and can be used to forecast the ice regime and support decision making such as on the need for artificial ice-breaking (by airforce bombing) and reservoir regulation (by YRCC). Moreover, using the forecasted air temperature data as input for running the model, this can clearly prolong the lead-time. During river freeze-up and breakup, the ice regime can be adequately predicted for supporting decision making on reservoir regulation and other measures, such as for Liujiaxia reservoir regulation to diminish the possibility of ice jams and ice dam occurrence, and Wanjiazhai reservoir regulation to create an artificial flood that flushes the Tongguan Heights.

Samenvatting (Netherlands)

IJsvorming op rivieren doet zich op veel plaatsen in de wereld voor, niet alleen in arctische maar ook in gematigd klimaatgebieden. Hoewel er relatief weinig aandacht aan wordt besteed in de hydrologische literatuur, kan ijsvorming op rivieren grote consequenties hebben waaronder het veroorzaken van overstromingen door kruiend ijs tegen dijken, het stagneren van scheepvaart, of verminderde productie van energie door waterkracht, en de invloed op de waterkwaliteit en het milieu. IJs heeft een belangrijke invloed op biochemische cycli en op de ontwikkeling van bijzondere ecosystemen. IJsvorming in rivieren omvat vele processen: de eerste formatie, evolutie, transport, accumulatie, dissipatie en dooivorming. IJsvorming in rivieren wordt bepaald door interacties tussen hydrodynamische, mechanische en thermische processen die tevens beïnvloed worden door meteorologische en hydrologische condities. IJsvorming in rivieren is van belang voor watermanagement in koude gebieden en is van invloed op het ontwerp en onderhoud van ondermeer reservoirs en waterwerken. Problemen die zich voordoen ten gevolge van ijsvorming zijn de vorming van ijsschotsen die dijken en dammen kunnen bedreigen, alsmede milieu effecten kunnen veroorzaken en van invloed zijn op morfologische processen.

De Ning-Meng reach (waarvan de Ningxia Hui en Inner Mongolië autonome gebieden deel uitmaken) is gelegen in het noordelijke deel van de Gele Rivier tegen de grens van China met Mongolië, met een lengte van 1237 km. Vanwege zijn specifieke geografische ligging en stromingsrichting bevriest de Ning-Meng reach iedere winter. Zowel tijdens bevriezen als ontdooien ontstaan omstandigheden die kunnen leiden tot het vastlopen van ijsschotsen en het ontstaan van ijsdammen met dijkbreuk en overstromingen als mogelijk gevolg, zoals vaak gebeurd in de geschiedenis van de Gele Rivier. Met de ontwikkeling van een integraal beheerplan voor de gehele rivier zijn de voorwaarden geschapen om veiligheid tegen overstromingen te bieden alsmede de beschikbaarheid van het schaarse water te garanderen. Dit plan vormt de basis voor het ontwikkelen van een gedetailleerd voorspellingsmodel voor bevriezen en ontdooien van de Ning-Meng reach. Daartoe is door de YRCC een computermodel ontwikkeld in aanvulling op veldmetingen en laboratoriumonderzoek, teneinde op die manier de onderhavige processen in de noordelijk tak van de Gele Rivier beter te begrijpen.

Hoewel numerieke ijsmodellen voor meerdere rivieren in de wereld zijn ontwikkeld, hebben deze veelal tot doel om te worden toegepast over korte afstanden. Bovendien zijn ze vaak niet ontwikkeld voor het doen van voorspellingen en missen ze het vermogen om te gaan met complexe stromingspatronen en rivier geometrieën. Ook geldt dat, vanwege de complexiteit van ijsvormingsprocessen, hydrodynamische en thermodynamische processen moeten worden gekoppeld waarbij de waterbalans (continuïteitsvergelijking) moet worden aangepast. De snelle economische ontwikkelingen en menselijke activiteiten hebben het gebied gedurende de laatste decennia verregaand veranderd. Vandaar dat het van groot belang is om ijsvorming en gevaar voor overstromingen tijdig te kunnen signaleren, waarbij computermodellen van groot belang zijn.

Tijdens dit onderzoek zijn hydrologische en meteorologische gegevens gebruikt uit de periode 1950-2010 om de karakteristieken van ijsvorming in de regio te bestuderen, met name nadat het Liujiaxia reservoir in gebruik was genomen in 1986. Sommige gegevens konden worden gebruikt om ijsvorming te modelleren en parameters te verifiëren die bepalend zijn voor correcte modelvorming. Daarnaast werden in het kader van dit onderzoek aanvullende meetgegevens verzameld voor verdere calibratie en validatie. Door het combineren van meteorologische en statistische voorspelmodellen voor de middellange en korte termijn kon een numeriek model worden ontwikkeld met verbeterde eigenschappen voor de voorspelling ijsvorming en mogelijke overstromingen.

Gedurende winterse omstandigheden in de Ning-Meng reach moet de continuïteitsvergelijking in het model worden aangepast door rekening te houden met ijsvorming en de opslag van water in uiterwaarden die van invloed zijn op de massa balans. Ook de impulsvergelijking behoeft aanpassing ten gevolge van toegenomen wrijving van de stroming tegen het ijsoppervlak. De hier voorgestelde aangepaste continuïteitsvergelijking verzekert niet alleen massabehoud, maar kan ook worden gebruikt voor aangepast reservoirbeheer tijdens de dooiperiode. De dikte van de ijslaag aan het oppervlak is gelijk aan een blok ijs dat aan het oppervlak drijft met een soortelijk gewicht van 0,9. Voor het modelleren van watertemperatuur en ijsvorming zijn in deze thesis vereenvoudigde formuleringen ontwikkeld die snelle berekeningen mogelijk maken. Er is een empirische relatie vastgesteld op basis van luchttemperatuur, dwarsdoorsnede van de rivier, afvoerdebiet, en ijsdikte waarmee het moment van bevriezen en ontdooien adequaat kan worden bepaald. In combinatie met 10-daagse temperatuurvoorspellingen kan daarmee een zeer effectieve voorspelling worden bereikt voor het voorspellen van vriezen en dooien op de middellange termijn.

De zogenaamde één-tegelijk methode is in deze thesis gebruikt voor een gevoeligheidsonderzoek naar de parameterwaarden voor ijsruwheid, ijsgroei, warmte uitwisseling met de atmosfeer, en watertemperatuur. De criteria voor het bepalen van het moment van bevriezen en ontdooien zijn gecalibreerd op basis van de waargenomen hydro-meteorologische omstandigheden. Daarmee werd een onzekerheidsanalyse uitgevoerd met als belangrijkste resultaat de waarden voor (i) waterdiepte en (ii) ijslaagdikte in Sanhuhekou station. Op basis van Monte Carlo simulaties werd de onzekerheid in parameterwaarden vastgesteld.

Samengevat kan worden geconcludeerd dat het numerieke ijsmodel dat hier is ontwikkeld voor de Ning-Meng reach goed kan worden toegepast voor het bepalen van de water temperatuur, ijsvorming, ijsdichtheid, ijsdikte, tijdstip van bevriezen en ontdooien van de rivier, afvoerdebiet, waterdiepte en wateropslag. De resultaten van het model lijken adequaat voor het doel: het voorspellen van de effecten van ijsvorming en het ondersteunen van te nemen maatregelen waaronder het doen breken (door middel van luchtmacht bombardementen) van ijsschotsen en het beheren van het waterpeil in de reservoirs (door de YRCC). De voorspelingstermijn kon aanzienlijk worden verlengd door gebruik te maken van verwachtingen van de luchttemperatuur als invoerwaarden voor het model. Het model is zeer geschikt voor reservoirbeheer met name ten tijde van bevriezing en ontdooiing, zoals het voorkomen van ijsophoping in het Liujiaxia reservoir, en het reguleren van het Wanjiazhai reservoir voor het doorspoelen op de Tongguan vlakte.

摘要 (Chinese)

在位于寒带或温带地区的许多河流中，结冰现象在一年当中经常发生。虽然在水文的教科书中很少涉及冰相关的专业知识，河冰对水文的影响却非常显著，包括由冰塞引起的极端洪水事件，影响航运和水力发电，冬季低流量及相应造成的生态和水质影响。冰是影响生物地球化学循环和相应生态系统的显著因子。

河冰现象包括各种冰的形成、发展、输移、聚积、消散、退化等。河冰过程是指在气象条件和水文条件的影响下，水动力学、动力学和热力学过程的复杂相互作用。在寒冷地区，河冰的存在是水资源开发中要考虑的重要现象。冰的相关信息可影响水库的设计、运行和维护。与河冰相关的重要水利工程需要考虑冰塞洪水、水库运用、调水和环境、生态和河道形态的影响。

宁蒙河段（包括宁夏回族自治区和内蒙古自治区河段）位于黄河流域北部，河长 1327 千米，每年冬天，宁蒙河段都要封冻，且在封河和开河期间，由于地理位置的特殊性及河流流向，在遇到不利条件时，冰坝或冰塞造成堤防决口和漫堤，在历史上形成巨大灾害和财产损失。随着黄河流域水资源统一管理的发展，在凌汛期间黄河上游的水量调度的要求不仅要满足凌汛期间的安全，而且还要充分利用有限的水资源。因此对冰情的预报在预报内容、时效和准确度方面提出了更高的要求。为解决上述提到的问题，黄河水利委员会提出要建立冰凌数学模型，用于弥补实地观测和实验研究的不足，也可进一步了解黄河河冰、冰凌形成的物理过程。

截至目前，在世界上有些河流已经建立了冰凌数学模型，它们主要以模拟为主，应用在较短的河道上。这些模型缺乏对复杂流动形态和河道情况的详细考虑。由于河冰现象在每个阶段的复杂性，需要根据不同的标准对水量平衡和热量平衡要分别进行计算，以求对水量进行平衡调整。随着社会经济的发展，以及气候变化和人类活动的影响，冰凌的一些特征发生了变化，特别是封河和开河期间造成的冰凌灾害。因此，对冰情进行分析，了解冰凌特征和利用数学模型进行冰凌预报、防止和减少冰凌灾害是非常重要的。

本次研究中，利用 1950 年到 2010 年的水文和气象数据，特别是 1986 年刘家峡水库投入运行后，对宁蒙河段的冰情进行了分析和总结。一些结论可用于河冰数学模型中的系数和参数的率定，另外，一些成果可用于解决冰凌模型建立中遇到的一些关键问题。而且，也开展了用于冰凌数学模型验证和率定的实地观测。

利用气象数值模型预报和统计预报联合的方法建立了宁蒙河段中短期气温预报模型，气温预报的成果作为冰凌数学模型的输入，可有效提高冰凌数学模型的预报预见期。

考虑到宁蒙河段的具体情况，在连续方程中，为使水量平衡，增加了槽蓄水增量项，包括冰盖项和滩地冰水项。在动量方程中，为使动量平衡，增加了冰盖摩擦力项。槽蓄水增量项的提出，不仅水量得到了平衡，而且提供了计算槽蓄水增量不同组成部分的水量，在开河期对水库调度具有重要参考依据。对于水面上冰层厚度等于水面上漂浮冰块厚度的问题，引入了淹没系数取 0.9。对于水温和流凌的模拟，提出简化的公式使计算更快捷。基于气温、河道形态、流量和冰厚，建立了河流封河和开河判断经验公式，可以判断河流是否封河和开河。利用验证的参数以及 1 到 10 天的气温预报成果，可以有效的确定河流是否封河和开河，且具有较长的预见期。

单个单次（One-At-A-Time）敏感测试方法用于模型的参数敏感性分析。利用实测的水文气象数据对冰盖糙率、冰盖衰减常数、水与大气的热交换系数、水温计算、河流封河和开河判断等参数进行了率定。在敏感性分析的基础上，将不确定性分析分为三湖河口水文站的水位和冰盖厚度不确定性分析两个部分，采用蒙特卡罗（Monte Carlo）模拟方法进行参数不确定性分析。

宁蒙河段冰凌数学模型可应用于模拟计算水温、流凌密度、冰盖厚度、河流封河和开河日期，流量、水位和槽蓄水增量，冰凌数学模型计算结果可信可靠。特别是将预报的气温成果作为冰凌数学模型的输入，可明显提高模型预报的预见期。预报的冰情信息可供人工破冰和水库调度等决策支持使用。例如刘家峡水库调度控制河流文封和文开，减少冰凌灾害发生几率；万家寨水库调度形成人造洪水冲刷潼关高程。

Contents

Chapter 1   Introduction

1.1 Background

The Yellow River, located in the Northern part of China, is the second longest river in China after the Yangtze River, and the sixth longest river in the world. Being the cradle of Chinese civilization and at the centre of China's current political, economic and social development, the river is known as the 'Mother River of China'. Four types of floods occur in the Yellow River basin: (i) summer floods; (ii) autumn floods; (iii) ice floods (winter) and (iv) peach floods (spring). The summer and autumn floods are mainly due to precipitation, and since the rainfall intensity in summer is much stronger than in autumn, the flood peaks of summer floods are larger but the duration shorter, compared with autumn floods. Ice floods occur in winter during the river freeze-up period due to ice jamming. Peach floods occur in spring during breakup of the frozen river due to ice jams and ice dam build-up. Since this is the time that the peach tree flowers, the flood is called peach flood.

Other high latitude rivers in the world encounter ice floods much less frequently. However, the special geographical location and difference in latitude of the Ning-Meng reach (including Ningxia Hui Nationality Autonomous Region and Inner Mongolia Autonomous Region) of the Yellow River, together with the river flow direction from South to North in that reach, lead to ice floods and peach floods. Since during the river freeze-up period, ice will occur in the downstream part earlier than in the upstream part, the downstream ice will block the water and cause ice frazil to form ice jams. And reversely, during the river breakup period the downstream part will de-freeze later than the upstream part, which easily causes ice dams or backwater to be formed which can result in ice flood disasters such as dams being destroyed or dikes breaching. In the past dike breaching resulted in huge casualties and property losses throughout history in the Ning-Meng reach of the Yellow River.

According to historical data from the Yellow River Conservancy Commission (YRCC), ice disasters have occurred frequently. Ice flood disasters occurred every year between 1855 and 1949. During that period ice floods destroyed dikes 27 times. Apart from that, there were 28 ice flood seasons with ice disaster from 1951 to 2005 (Rao et al., 2012). After the operation of Liujiaxia Reservoir in 1968 and the Longyangxia Reservoir in 1986, ice flood disasters only occurred in 1993, 1996, 2003, and 2008 (Gao et al., 2012). Taking the most recent one as an example, during the winter season 2007/2008 the channel water storage in the Inner Mongolia reach attained the largest value of 1.835 billion m^3, which is 0.595 billion m^3 more, compared with normal condition (1.240 billion m^3). A map of the main ice flood disasters on the Ning-Meng reach is presented in Figure 1.1.

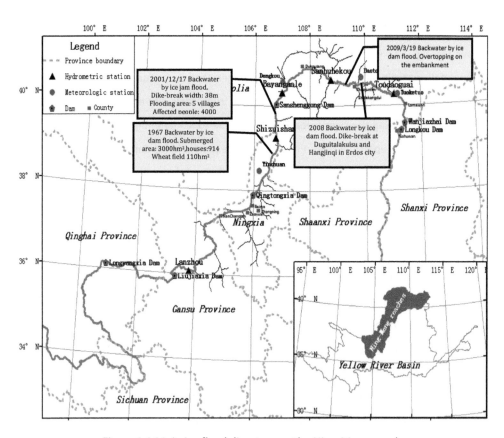

Figure 1.1 Main ice flood disasters on the Ning-Meng reach

As a result, the water level (using the Huanghai sea level as datum) at Sanhuhekou hydrological station reached a level of 1,021.22 m which was 0.41 m higher than the highest-ever level on record of the station, leading to dike-breaking at two sites in Duguitalakuisu, Hangjinqi County (Figure 1.2 and Figure 1.3) in Erdos City, causing serious economic losses (Chen et al., 2012). Based on literature review, the main ice flood disasters that occurred in the Ning-Meng reach are presented in Figure 1.1. Especially in the Ning-Meng reach, ice jams, ice dams and other ice flood disasters happened frequently. Statistics of ice flood disasters in the Ning-Meng reach during the period from 1901 to 1949 are shown in Table 1.1.

Table 1.1 Statistics of ice flood disasters on the Ning-Meng reach (1901-1949)

Year	Ice flood disasters
1901	During the breakup period, ice run resulted in ice dam, and then ice flood occurred to get the livestock and grassland submerged.
1910	During the breakup period, ice run resulted in ice dam, and then ice flood occurred to get the livestock and grassland submerged.
1926	Among the Sanshenggong reach, ice flood caused by ice jam occurred and water level reached the roof.
1927	In March, along the Yongqi channel at Linhe, ice flood leaded to dike-break, about 400 houses were submerged, financial loss: RMB 200,000.
1929	The water level increased due to ice jam along the Yongqi channel at Linhe, as a result, the Lisan ferry breached.
1930	On March 21, ice flood submerged land with 1,000 arces along the Yongqi channel at Linhe.
1932	In spring, ice flood leaded to dike-break.
1933	In March, at Dengkou in the east of Sanshenggong, dike-break resulted in a flood area of 150 km^2.
1935	At Qingtongxia, Majiatan ice flooded area reached about 20 km^2, dozens of houses were destroyed, 2 children, 4 cows died.
1936	In spring, at Dengkou lots of residents and livestock were submerged, and from March to May, traffic was interrupted.
1945	In spring, ice jam occurred at Takouwan of Linhe, and then the county was flooded.
1947	In mid-March, ice broken up and resulted in dike-break at Dengkou.
1949	In March, ice jam occurred at Dengkou and resulted in ice flood.

Figure 1.2 Dike breach in Duguitalakuisu County in 2008

Figure1.3 Dike restoration in Duguitalakuisu County in 2008

It can be seen that on the Yellow River the main problem about ice floods is caused by ice dam formation and ice jamming which could result in dike breaching and overtopping of the embankment. Furthermore, non-engineering measures such as ice regime information observation and ice regime forecasting and early warning are very important for decision makers to take engineering measures or other measures such as bombing by airplanes or artillery to control ice flood hazards. This is a key problem to be solved in the Ning-Meng reach of the Yellow River basin.

1.2 Problem statement

Based on several years of ice flood control experience in the Ning-Meng reach during the river freeze-up and breakup period, it was considered better to reduce the discharge in advance through upstream reservoirs regulation, in order to decrease the opportunity for ice jams and ice dams to develop. However, reducing the discharge of the reservoirs will decrease the electricity power generation, which will influence human living conditions and production capabilities of industry etc. Therefore, in order to diminish this effect, the proper discharge should be determined according to the real ice regime conditions and channel conditions. Especially and most importantly, the date for river freeze-up and breakup should be forecasted as accurately as possible in order to support proper reservoir regulation. For the Ning-Meng reach, the reservoirs that can be used to regulate ice flood control are the Longyangxia reservoir and Liujiaxia reservoir which are located upstream at a distance of at least 779 km. For a flow propagation time between 7 to 23 days, this means that if one wants to control the discharge at the Ning-Meng reach using the upstream reservoirs, one should know the ice regime conditions with a lead-time of at least 7 days. This puts high requirements on the ice regime forecasting accuracy and lead-time precision. Following the development of an integrated water resources management approach for the Yellow River, the requirements on water regulation of the upper Yellow River during ice flood period are not only to safeguard against ice floods, but also to satisfy and sufficiently utilize the limited water resources. Since 2006, the Yellow River Conservancy Commission (YRCC) put forward procedures to utilize the peach flood to flush sediment and decrease the Tongguan Heights (Tongguan is located at the junction of Shanxi, Henan and Shaanxi province, which contains the control section of the Yellow River for water and sediment; upstream is the entrance of the Weihe River flow into the Yellow River, downstream is the Sanmenxia reservoir). This requires more information such as when the flow peak will occur and what the peak volume release procedure will be to regulate the Wanjiazhai reservoir, which is located at the downstream end of the Ning-Meng reach, to create an artificial flood to flush the sediment and decrease the sediment at the Tongguan Heights. Ice regime information gathering and forecasting are the most important scientific support measures for ice flood control, reservoir regulation and decision-making. Although models are only simplified representations of the real world (Mynett, 2002; Price, 2006), the YRCC decided to build a numerical ice flood model coupled with medium range air temperature forecast model for the Ning-Meng reach in order to simulate and predict ice regime processes during the ice flood period. This thesis research is part of that process. Although, there are several reasons and necessities for building a numerical ice flood model for the Ning-Meng reach, first the characteristics of ice regime variations should be studied and explore whether the present ice regime forecasting methods can satisfy the requirements of ice flood control as well as water regulation. In recent years, with the influence of human activities and effects of climate change, the mean air temperature increases gradually with large variations in air temperature in winter, which caused river reach freeze-up and breakup to occur 2 to 4 times during one ice flood period. With the runoff decreased in the upper Yellow River and more water used in the irrigated area in the Ning-Meng reach, together with the Longyangxia reservoir that was put into operation in 1986, the runoff of the Ning-Meng reach has changed due to the regulation of the reservoirs, leading to lower runoff causing more sediment, river channel siltation and river bed elevation increase as well as river flow capacity decrease. Any water level increase during the freeze-up and breakup period of the ice flood season, with the low standard dikes in the Ning-Meng reach, very easily can cause ice disasters such as dike breaching. When the river is frozen, it acts as a closed conduit, and the discharge capacity is known to depend on the overall roughness. Due to variations in air temperature, discharge and river channel, the channel water storage obviously increased since 1990. Especially in the ice flood season 2004/2005, the maximum channel water storage reached over 1.91 billion m^3, while the annual mean maximum since 1950 is about 1.24 billion m^3. Channel water storage and its spatial and temporal distribution of growth and release procedures should be studied for the purpose of utilizing the peach flood to flush the Tongguan Heights. The present ice regime forecasting procedures including ice run date, freeze-up and breakup date, freeze-up and breakup water level and discharge are based mainly on statistical forecasting schemes. These cannot provide information about the spatial and temporal distribution of the river freeze-up, ice depth, frazil jam, ice jam, channel water storage and variation in river ice processes. Hence YRCC cannot satisfy the actual requirements of ice flood control and water resources utilization based on statistical methods only, and it is necessary to build up a numerical ice flood model for the Ning-Meng reach following the considerations mentioned above.

1.3 Objectives and research questions

1.3.1 Objectives

Already several one-dimensional and two-dimensional numerical ice regime models have been built for various rivers around the world (Lal and Shen, 1981; Beltaos, 1983; Shen, Wang, Lao, 1995). However, these models usually only have simulation purposes for short river courses and are not set up to make any forecasts. The level of research activity on river ice has been much less than research with ice-free conditions. Significant progress has been made in the last couple decades but much work still needs to be done (Beltaos, 2008; Shen, 2010). Ice flood models have been developed for one-dimensional steady state conditions, one-dimensional unsteady state conditions, two-dimensional spatial models and data-driven models that can also be applied to forecast ice floods. In order to build up a numerical ice flood model for the Ning-Meng reach with a distance of 1,237 km, several difficulties should be solved, such as how to account for river ice transport processes including the transport of thermal energy, how to deal with undercover transport of frazil granules, surface ice transport, ice cover friction, jam formation, growth and release of channel water storage and its spatiotemporal components and distribution, how to develop river freeze-up and breakup criteria, how to couple meteorological models and ice models, etc.

This research focuses on the Ning-Meng reach of the Yellow River basin in China, where ice flood problems require increasingly more attention from YRCC and from the China National

Flood Control and Draught Relief Headquarters Office, and more and more information needs to be provided for decision making during the ice flood period. The principal aim of this research is (i) to analyse ice regime characteristics of the Ning-Meng reach over the recent 50 years, especially after the Longyangxia reservoir was put into operation in 1986 which changed the river flow from a natural runoff to a reservoir-regulated river; and (ii) to build a generalized numerical ice regime model for decision making on ice flood control and water resources regulation of the Yellow River.

1.3.2 Research questions

The specific research questions that guide this research are:

Question 1: Can we make specific observations on ice regime to improve the representation and parameter selection of river ice processes and hydraulics?

Question 2: Can we determine the effect of river ice cover friction on the river flow, and use the in situ observation data on actual ice regime conditions to quantify the ice cover roughness and decay parameters for different hydrometeorological conditions in order to reflect the actual situation?

Question 3: Can we identify the components and formation mechanisms of channel water storage, and how to simulate its spatial and temporal distribution, as well as ice growth and release processes?

Question 4: Can we determine the river ice freezing and thawing criteria based on thermal, mechanical, and river course conditions?

Question 5: Can we couple a numerical meteorological model with a numerical ice flood model, in order to prolong the lead-time of ice regime forecasting and provide early warning?

Question 6: Can we build a suitable and applicable numerical ice flood model for the Ning-Meng reach of the Yellow River that contains proper simulation of the ice regime and has ice flood warning and decision support functions?

1.4 Thesis Outline

This thesis is composed of six chapters.

Chapter 1 gives the background, problem statement and objectives of this research.

Chapter 2 addresses the state-of-the-art theoretical knowledge and the available modelling techniques about ice regimes and ice floods. This includes summarizing the dominant river ice processes, classifying ice flood formation, exploring ice model availabilities and limitations, and identifying what aspects need improvement in case of one-dimensional ice flood modelling.

Chapter 3 focuses on the study area, data availability, field campaign measurements including data analysis of ice regime characteristics, and data used for model setup, parameter calibration and verification. The characteristics of the ice regime in the Ning-Meng reach are analysed and the dominant factors for the ice regime in the region are established.

Chapter 4 is the key part of this thesis, presenting the one-dimensional ice flood model structure and mathematical representation, especially the mathematical formulations that reflect the typical behaviour of the Ning-Meng reach during river freeze-up and breakup time. A modelling framework is proposed for two-dimensional river ice flood modelling and further research. The combination with a numerical meteorological forecast model is described and a statistical method to set up medium and short-range air temperature forecast models are outlined. Parameters and verification results of ice flood modelling for the Ning-Meng reach test case are presented and discussed.

Chapter 5 describes a sensitivity analysis and uncertainty analysis of the established model, and shows the ice flood modelling results used for ice flood control by the headquarters office of YRCC.

Chapter 6 gives the conclusions and recommendations of this research.

Chapter 2 Ice Flood Processes and Models

River ice is a natural phenomenon, which could be commonly seen in the cold regions of the world. As summarized by Sun and Sui (1990), 82% area in North America, which includes the whole area in Canada and 52% area in USA; the majority of regions in Russia; Norway, Finland, Sweden in North Europe; and China and Japan in Asia; are the regions where there are river ice problems present. River ice plays an important role in the regions in the northern hemisphere of the earth, which could be divided into positive and negative effects by Hicks et al. (2008). On the positive side it can be mentioned that in winter the main transportation means are ice roads and ice bridges caused by river ice in the northern regions of Canada, Russia, USA (Alaska) with sparse population. However, on the negative side, river ice could cause ice flooding, hamper hydropower generation, threaten hydraulic structures, hinder water supply and river navigation and other aspects.

Many rivers in the world are experiencing ice floods. For example, Hay River in Canada, Vistula River in Polish, Karasjok River in Norway, Red River in America, and Yellow River in China are the main rivers suffered from ice floods in the world. The Hay River in the Northwest Territories of Canada is a river where the ice jam flooding happens frequently, especially mechanical breakup during the breakup period, although the thermal breakup occurs sometimes. In the majority of cases the mechanical breakup happened which resulted in the risk of ice flood, as it is the cases from 1964 to 2008, when in 9 years Hay River suffered from the significant ice floods and other 9 years the Hay River suffered from moderate ice floods (Kovachis et al., 2010).

Vistula River is the largest river in Poland, with a total length of 1,047 km. Vistula's flow direction is from the south to the north, and discharges into the Baltic Sea. The river suffered from ice formation, especially on the final section of Lower Vistula with a total length of 390 km. Most floods occurred on this last section are due to ice jams. For example, in 1982, severe floods happened in the region of upper part of Wloclawek reservoir, which were caused by large ice jam (Majewski and Mrozinski, 2010). Norwegian rivers experience ice jams every year, which result in inundation and damage on infrastructure. For example there were three floods which were induced by ice in 1917, 1932, and 1959 on Karasjok River (Lier, 2002).

Along the Red River near Netley Cut, which originates from South Dakota in United States and flows north to Lake Winnipeg, there was a lot of severe ice floods happened in history. For example in Manitoba, Canada, the most severe ice jams and floods happened north of Winnipeg which is located between Selkirk and Lake Winnipeg (Haresign and Clark, 2011). River ice could result in the formation of ice cover at the downstream of a dam, which could lead to the decrease of the water level difference between the upstream and downstream of the dam, hence the hydropower generation efficiency could decrease. Apart from that, the frazil ice and hanging dams' problems could threaten the safety of hydropower structures. Two such examples are the Nelson River in Canada and Kemijoki River in Finland.

The Nelson River is in the northern Manitoba, Canada, where three largest hydropower stations in Manitoba Hydro's system on this part of river are located. In winter, open water with large areas gave enough time to form and evolve large quantities of frazil, anchor ice, and surface ice. it happened very often that an ice cover would form on the lower reach of the river initially, and then propagate upward to the existed stations because the surface and suspended ice accumulated, which could result in the decrease of the difference of water level between the upstream and downstream of the dam, and further the hydropower generation efficiency could decrease. According to the statistics, the financial and operational implications to the river ice processes are about one million dollars per year (Malenchak et al., 2008).

The Kemijoki River is the longest river in Finland, whose basin covers a large area of Northern Finland. And there are 16 hydropower stations on the river to produce hydropower. However, the river is prone to frazil ice problems due to hanging dams, which could be harmful to hydropower production and environment (Aaltonen et al., 2008).

River ice could hinder water supply and river navigation. In winter, frazil ice could be formed which could result in frazil blockage of intake screens, and further lead to hindering water supply. An example is given by Altberg (1936) about the city of St Petersburg, in Russia, where the whole water supply system was paralyzed by frazil blockage for three days. And in Lake Michigan nine frazil blockage events occurred during the winter of 2006/2007 (Daly and Ettema, 2006). The formed ice could threaten the safety of ships, such as border ice, ice cover, and anchor ice. For example the Sayano-Shushenskoye reservoir, which is the deepest one located in Russia, if there is no ice phenomenon, without doubt it could be used for navigation. However, because of the ice phenomenon the possibility of navigation should be checked (Kolosov and Vasiljevskiy, 2004).

On the Yellow River, in China, the ice disaster appeared frequently. According to historical data of Yellow River, there were ice flood disasters every year between 1855 and 1949, during the period the dikes were destroyed by ice floods for 27 times. Apart from that, there were 28 ice flood seasons with ice disaster from 1951 to 2005 (Rao et al., 2012).

2.1 River ice flood processes

As mentioned above, river ice could cause ice flooding, hamper hydropower generation, threaten hydraulic structures, and hinder water supply and river navigation. The classification, definition, and formation mechanism of ice flood is presented below as it was introduced by Liu et al. (2000).

2.1.1 Ice flood classification and definition

According to the formation causes of ice flood, ice flood of rivers could be classified as ice jam flood, ice dam flood, and ice-snow melt flood. Ice jam flood is happening when a lot of frazil ices and crushed ices gather under the ice cover, which could lead to the increase of water level on the upper reach. After accumulation of frazil ices crushed ices is formed that will develop toward upper reach and also progress slowly towards the lower part of the reach. Both ice accumulation and backwater can cause the ice jam. When the backwater level exceeds the critical height above which destroying dams and dike-break could happen, ice jam flood occur.

Ice dam flood is happening when a lot of fluid ices dive, press, and accumulate, which could cause the increase of water level on the upstream part of a blocking section of a reach. Huge water-blocking ice deposit formed by diving, pressing and accumulation of ice is like an ice dam. Therefore, both backwater caused by ice dams and the ice dam body itself are called ice dam. When the backwater level exceeds the critical height above which destroying dams and dike-break could happen, ice dam flood occur. Apart from that, when the external forces acted on the ice dam along the flow direction exceeds the internal forces, ice dam could collapse. Just like dam breakup, the water with ice would outburst, which could also result in ice dam flood.

Ice-snow melt flood is happening during the breakup period in spring, channel storage increment (river network or river channel storage of ice-snow, freezing water quantity in soil and rainfall in melting period and flow inverting from snow) is released to form flood, the flood is named as ice-snow melt flood.

2.1.2 Ice flood formation mechanism

The ice flood formation mechanisms and comparison are shown in Table 2.1 and explained below.

Ice jam flood

At the initial stage of winter, since the air temperature decreases, the heat lost exceeds the amount gained on the surface of water body, the water temperature may decrease to the freezing point; further heat loss could result in super cooling phenomena and the formation of ice crystals. Based on the formation of frazil ice, the frazil ice at the bottom of the river channel may attach on the river bed and objects in the water, which could result in the formation of anchor ice. With the increase of the size and amount of frazil ice, buoyancy could exceed vertical forces, and then it could float on the surface of water and lead to high ice run concentration (density), after thermal thickening, the ice pans could form (Daly, 1984). The ice-cover period starts from the formation of border ice; it could occur due to thermal growth, apart from that it could laterally grow along the existing border ice because of the accumulation of surface ice. The Figure 2.1 illustrates the ice cover/ice jam evolution and formation.

The increase of surface ice run could cause the stoppage on the surface of water, and then the ice cover starts to occur. Once the ice cover forms, the incoming ice from the upstream could accumulate to make the ice cover extend to the upstream. When the flow velocity exceeds a critical value, the ice cover will stop extending, the water surface ice reaching the leading edge of ice cover will submerge under the ice cover and continue to transport downward. With the increase of ice cover thickness, the ice jam occurred; the water level at the upstream of ice jam could increase, when the water level increase to surpass the height of embankments, the ice flood could occur.

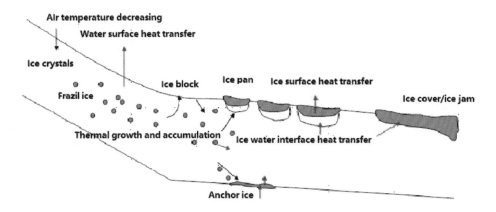

Figure 2.1 Ice cover/ice jam evolution and formation

Ice dam flood

During the breakup period, ice dams could form due to enough flowing ice and suitable channel topography condition, such as continual bend channel, narrow section and so on. Once the ice dam happens, similar to the ice flood caused by ice jam, the water level at the upstream of ice dam could increase, and when it increases to some extent, it could result in ice flood. Apart from that, due to thermodynamic and hydrodynamic factors, when the ice dam increases to some extent, the external forces acted on the ice dam along the flow direction exceeds the internal forces, ice dam could collapse, just like dam breakup, the water with ice would outburst, which could result in ice flood.

Table 2.1 Comparison of the three types of ice floods

Item	Ice jam flood	Ice dam flood	Ice-snow melt flood
Occurrence time	At the initial stage of freeze-up period and after freezing-up.	During the breakup period.	During the breakup period.
Location	Mostly at river bends and places with the slope changing from being precipitous to being gentle.	Generally, the same as the ice jam flood, and also at the front edge of ice cover and ice-blocking area.	Frozen reach.
Air temperature features	Steep air temperature drop and continuous low temperature.	Air temperature rising to freezing point suddenly and then steep temperature drop, steep fluctuation range.	Air temperature rising to freezing point gradually and small fluctuation range.
Backwater height	Dependent on incoming water quantity, frazil ice quantity from the upper reach, and blocking degree of cross-sections. Higher backwater could occur on river course with valley type.	Dependent on incoming water quantity, ice quantity from the upper reach, hardness of ice quality, and blocking degree of cross-section. Mostly backwaters are high.	Dependent on incoming water quantity from the upper reach and ice run blocking degree, higher than that with the same discharge under smooth condition; while lower than that caused by ice jam and ice dam.
Channel storage	Large channel storage increment.	Channel storage increment released by reach is accumulated and stored in upper reach of the ice dam.	Decrease of channel storage increment along the river.
Ice flood peak	No distinct flood peak.	Large ice flood peak may occur before the formation of ice dam and large ice flood peak of discharge may form after the outburst of ice dam.	Distinct ice flood peak occurs and grows along the river.
Evolution features	Including the three stages: formation, steadiness and melting.	Including the three stages: formation, steadiness and outburst.	Including courses of melting, ice flood peak, high water level (rising difference) swelling and disappearance.
Outburst situation	Slow outburst.	Abrupt outburst by hydrodynamic action or human activities.	After moving toward lower reach, the river course gets back to normal condition.
Ice flood disaster	Partially severe flood damage.	Partially great flood damage and ice run damages to water conservancy facilities.	Flood damage to some extent, but usually light.

Ice-snow melt flood

During the breakup period in spring, the air temperature increases to above freezing point, ice and snow formed by river network or river channel storage of ice-snow, freezing water quantity in soil and rainfall in melting period and flow inverting from snow starts to melt, which could result in the increase of water level and discharge during the breakup period. When the backwater level exceeds the critical height above which destroying dams and dike-break could happen, ice-snow melt flood could occur.

2.2 River ice research

River ice phenomena include formation, evolution, transport, accumulation, dissipation, and deterioration of different kinds of ice. These phenomena include complicated mutual effect among river hydrodynamic and thermodynamic processes, which are under the influence of meteorological and hydrological conditions. The river ice processes can be divided into three main periods, namely the freeze-up period, ice-covered period, and breakup period. As summarized by Shen (2006), in the past fifty years, the engineering and environmental issues have largely driven the development of river ice research, and significant achievements have been made during this period of time. However, Beltaos (2008) thought that although there was remarkable progress in understanding and quantifying the complex river ice processes, yet, many problems about river ice still remained unsolved or partially solved. According to the summarization of Shen (2006), the river ice research areas could be divided into the following two parts:

- Energy budget methods and water temperature distribution calculations before and during the freeze-up period are well understood e.g. Shen and Chiang (1984);
- The mechanisms of supercooling and frazil ice formation, which are also relatively well understood (Osterkemp, 1978; Daly, 1984).

The evolution of frazil ice, frazil floc, and anchor ice and anchor ice dams needs to be studied (Ye et al., 2004). The transitional conditions among different ice run regimes are not understood clearly (Hammar et al., 2002). The knowledge on the mechanism of ice pan and ice floe formation is limited. A complete analytical formulation about the mechanical breakup needs to be developed. At the same time, Beltaos (2008) showed the challenges and opportunities in the research of river ice processes. The main challenge is to avoid or decrease the negative influence of the river ice processes and to make sure that the positive influence is not affected by the human activities. In order to meet the challenge, it is necessary that there is a good understanding of river ice processes qualitatively. However, there still remain serious gaps. Such as the research about anchor ice, breakup, ice jamming, and climate change influence on the river ice process. The main opportunities are how to use the new technologies to understand the river ice processes. New technologies include instrumentation, numerical modelling, mitigation and prediction of climate impacts on river ice processes.

2.2.1 Freeze-up period

In winter, with the decrease of air temperature, the water temperature will drop to the freezing point because of the heat exchange between the water body and the boundary. A number of researchers have tried to compute the heat exchange rate between the water body and the boundary. They are all based on the energy budget methods and empirical equations, such as detailed energy budget methods (e.g. Paily et al., 1974; Ashton, 1986; Hicks et al., 2008). However, in practice the meteorological data are not easily available. Hence, approximated linear relationships (Paily et al., 1974; Ashton, 1986) are deduced, which also includes formulas of the heat exchange rate at the water-ice interface were provided by Ashton (1979). The method to compute the heat exchange at the water-river bed interface was developed by O'Neil and Ashton (1981).

When the water temperature drops to the freezing point, further heat loss could result in the formation of skim ice and frazil ice. There is not too much work done on this topic, in history only Matousek (1984) put forward a semi-empirical method to determine the formation of different types of ice. The advantage of this method is that the data needed is limited. Based on the research from Matousek (1984), Lal and Shen (1991) found out that all the parameters related to the method should be calibrated according to the different rivers. Because the method proposed by Matousek (1984) is semi-empirical, future improvement about the method is needed. However, the method is still used until now, even applied in the mathematical model (such as RICE) to simulate the skim ice and frazil ice formation. There are a few researchers to do the research about anchor ice. And the understanding about anchor ice is still very limited.

In the last several years, a few statistical and experimental methods were used to analyze anchor ice. It is commonly known that the anchor ice could be an evolutionary form of frazil ice. These studies just focused on finding the relationships between some parameters which are easily calculated (such as Froude Number and Reynolds Number) and anchor ice formation (Doering et al., 2001; Hirayama et al., 1997; Kerr et al., 2002; Qu and Doering, 2007; Terada et al., 1998). In the same time, some research on analytical formulas is done (Tsang and Lau, 1995; Hammar et al., 1996; Yamazaki et al., 1996; Qiu, 2006). Shen and Wang (1995) put forward analytical formulas to depict the frazil ice evolution, which are used nowadays as a module in the RICEN model.

However, based on the field observations of anchor ice formation on the reach of the Laramie River, Kempema et al. (2008) argued that the anchor ice could be more than an evolutionary form of frazil ice; it is a major contributor to ice-cover formation in shallow rivers. Apart from that, a comprehensive set of parameters should be taken into consideration instead of the Froude Number and Reynolds Number, such as cooling rate and ice growth rate and so on. Hence, the research about anchor ice is rudimentary, but good progress in the last few days have been made by the lab and field measurement (Beltaos, 2008).

2.2.2 Ice-cover period

The ice-cover period starts from the formation of border ice. Border ice can form and grow in static mode and dynamic mode. In static mode, it grows from the skim ice, and in dynamic mode, it is due to the accumulation of surface ice pieces along the river bank or existing border ice. The research about border ice is mainly based on empirical formulas, such as Matousek (1984) and Miles (1993). The specific formula of Matousek could be seen in the model introduction part. Based on the empirical formulas, further analytical research with field data has been continued, Matousek (1990), Shen and Van DeValk (1984) found that when the averaged velocity u in depth was above a critical value 0.4 m/s, the border ice growth stopped. Svensson et al. (1989) put forward a critical value to limit the formation of static border ice; the critical value is depth-averaged water temperature beyond which border ice will not grow.

The increase of surface ice run could cause stoppage on the surface of water, and then the ice cover starts to occur, and as more and more ice produced, finally it could result in the ice cover jamming or bridging across the river. The research of ice cover has developed from static ice cover research to dynamics ice cover research which could be divided into one dimensional dynamics ice cover research and two dimensional dynamics ice cover research. Pariset and Hausser (1961) put forward an accumulation theory about water surface ice. According to the static balance of internal and external forces on the floating ice block, they deduced the formulas to determine the final thickness of water surface ice. Based on their work, a lot of researchers continued to refine and extend the accumulation theory, such as Uzuner and Kennedy (1976); Beltaos (1983); Beltaos and Wong (1986).

The above presented research is based on the static ice cover theory. However, the ice dynamics were not taken into consideration, Hence the static ice cover theory could not determine the time when and the place where the ice cover could occur. Based on the limitation, one-dimensional formulas for river ice transportation by flow constitutive laws were developed (Shen et al., 1990); Shen developed a one-dimensional model to simulate the dynamic ice transportation by MacCormack method. Due to the frictional resistance of riverbanks and channel bottom, irregular cross-section of the river channel, and unsteady flow state in reality, one-dimensional model could not meet the need to simulate the ice transport dynamics perfectly. Hence, a two-dimensional model named DynaRICE to simulate the ice transportation dynamics based on the Lagrangian Smoothed Particle Hydrodynamics (SPH) method was developed (Shen et al., 2000). The model DynaRICE will be discussed in the ice model section.

If the value of flow velocity is high enough, the progression of ice cover would stop, the incoming surface ice from the upstream at the leading edge will continue to transport under the ice cover, as the effective size of ice particles grows with the time, the buoyancy becomes large enough to create an upward movement, and then the ice particles would be brought to the underside of ice cover and deposited, and finally the frazil jam could form. The frazil jam is also named as hanging dam (Shen and Wang, 1995).

The research about ice transportation under the ice cover and ice jams started from a critical velocity criterion or Froude Number criterion (Kivislid, 1959; Tesaker, 1975), which could be used to determine the location and thickness. However, the method could not provide a way to compute the accurate value of critical velocity and Froude Number. Hence, Shen and Wang (1995) develop a concept of ice transport capacity to determine the ice deposition under the ice cover, If the local flow velocity reaches a critical velocity, the undercover deposition will cease, if the local velocity exceeds the critical velocity, the frazil ice starts to erode, and the ice transport capacity theory is demonstrated by the field observation data.

Although the ice transport capacity theory has been demonstrated by the field observation data, Beltaos (2008) still thought that the ice transportation under the ice cover was partially understood. As the heat loss of ice cover surface continues, the water during the interspaces of ice cover will freeze up from the water surface downward, which could lead to the increase of ice cover thickness.

Nowadays, it is relatively clear to understand the process of the thermal growth and erosion of ice cover. Firstly the research focused on the ice cover without layers of snow ice, snow slush and black ice, it is easy to use the simply thermal conduction equations to analyze the process (Ashton, 1986; Shen and Chiang, 1984).

Secondly, the layers of snow ice, snow slush and black ice has been taken into consideration when simulating the thermal growth and erosion of ice cover, the undercover deposition of frazil ice and the accumulation above the ice cover by snow could affect the growth and erosion of ice cover (Calkins, 1979; Shen and Lal, 1986).

Finally based on the previous research, a lot of models were produced to simulate the process. The most classical one was degree-day method (Stefan, 1889), which has been used to simulate ice cover growth for a long time. The shortcoming of the method is that it could only simulate the thermal growth of ice cover instead of both thermal growth and erosion of ice cover, and the parameter in the formula should be determined by the historical data at each case. Hence, Shen and Yapa (1985) developed a modified degree-day method after refining the classical degree-day method, in the modified method the decay of ice cover could be also simulated. A comprehensive model in which the heat exchanges among all the interfaces are taken into consideration was developed by Shen. The model was tested on St. Lawrence River in Canada, and the agreement between the observation data and simulation data was good.

2.2.3 Breakup period

When it comes to the arrival of spring, discharge and water level start to increase, and the ice cover begins to become weak, until the discharge and water level reach a critical point and the ice cover could be moved. Once in motion, if the discharge increases before the thermal melt-out, the mechanical breakup could occur; otherwise, the thermal breakup could occur. A mechanical breakup could result in severe ice run and ice jam, even ice flood, which could exert a negative influence on the hydraulic facilities and the safety of people living along the river. Hence, it is significant to understand the physical processes and simulate and forecast the mechanical breakup. However, the ability to simulate the mechanical breakup is still limited (Shen, 2006). A few researchers have tried to simulate propagation of ice jam release waves, most of them using the one dimensional model without the consideration of ice effect (e.g. Beltaos and Krishnappan, 1982; Blackburn and Hicks, 2003). Although the deviation between the observed data and simulated data was acceptable, it is difficult to match the shapes of observed stage hydrographs with those of simulated ones, which implied that the ice effect on the wave propagation could not be neglected reasonably. The field investigations were conducted by Jasek (2003), who found that release wave celerity seemed to change with different ice conditions. Based on the previous research output, Liu and Shen (2004) started to take the ice effect on the wave propagation into consideration, be coupling the flow dynamics and ice dynamics model to analyze the ice resistance effect on the wave propagation. Finally they found out that the ice resistance effect could decrease the peak discharge and propagation velocity. When it comes to the arrival of spring, discharge and water level start to increase and the ice cover begins to become weak, until the discharge and water level reach a critical point, the ice cover could be moved. Once in motion, if the discharge increases before the thermal melt-out, the mechanical breakup could occur; otherwise, the thermal breakup could occur.

She and Hicks (2006) built a model named River-1D about ice jam release waves with the consideration of ice effects. In the model three equations were included, namely the total (ice and water) mass and momentum conservation equations and ice mass conservation equation. In the total (ice and water) momentum conservation equation, an empirical term was used to depict the ice resistance effect; and in the ice mass conservation equation, an ice diffusion term was used to consider the ice effect. The model was tested by the release event on the Saint John River (Canada) in 1993, and 2002; and on the Athabasca River (Canada) in 2002. The test result was quite good. In addition to the content summarized above, statistical methods and ANN are also used to forecast the river ice breakup. The potential for using the Fuzzy Expert Systems to forecast the potential risk of ice jam was discussed (Mahabir et al., 2002), in the case study, antecedent basin moisture, late winter snowpack conditions, and late winter ice thickness data were used as training inputs to develop the Fuzzy Expert System. The Fuzzy Expert System was tested by the observed ice jam flooding event on Athabasca River at Fort McMurray from 1977 to 1999, and the system identified five years when the high water levels could occur, including four years when the high water level happened. Based on the potential research, more research has been done (Chen and Ji, 2005; Mahabir et al., 2006; Wang et al., 2008), they just took more related parameters into consideration.

2.3 Ice modelling

Among all the river ice research, mathematical modelling is an essential part of the progress. Mathematical models have been developed for different kinds of river ice processes. They could supplement inadequacies of field and lab studies, helping to understand the physical processes of river ice, at the same time. They could be a tool to help design and plan engineering projects. The main available ice models are summarized in Table 2.2, and at the same time, the model name, explorer, development time, main functions, and field application are introduced.

2.3.1 Ice model classification

The available models could be divided into 3 types (Debolskaya, 2009): 1D model, 2Dmodel and Data driven model (Figure 2.2):

- 1D model: HEC-2, ICETHK, HEC-RAS, RICE, RICEN and CRISSP1D.
- 2D model: DynaRICE, CRISSP2D.
- Data Driven Model: Fuzzy logic model especially Artificial Neural Network model (ANN).

Depending on the different flow states, 1D models could be divided into 2 classes (Shen, 2010):

- 1D model with flow of steady state: HEC-2, ICETHK, HEC-RAS.
- 1D model with flow of unsteady state: RICE, RICEN and CRISSP1D.

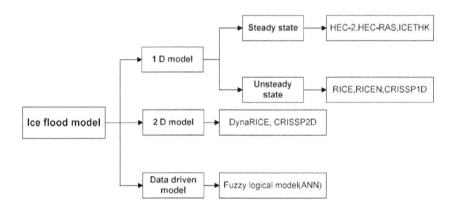

Figure 2.2 Ice model classification

Apart from the above classification, there is another one which is according to the different simulation functions, the ice models could be divided into 3 types (Shen, 2010): Ice jam model, Comprehensive model, Data driven model (Figure 2.3). Here, a comprehensive model is the combination of mathematical equations together with empirical formula to simulate the river ice phenomena.

- Ice jam model: HEC-2, ICETHK, HEC-RAS, DynaRICE.

- Comprehensive model: RICE, RICEN, CRISSP1D, CRISSP2D.

- Data driven model: Fuzzy logic model especially Artificial Neural Network model (ANN)

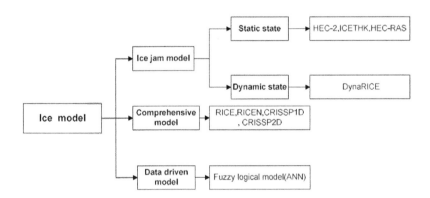

Figure 2.3 Ice model classifications based on different simulation functions

Table 2.2 Main ice models

No.	Model name	Explorer	Time	Main functions	Field application
1	HEC-2	U.S. Army	1990	The model could calculate water surface profiles under steady gradually varied flow condition.	-
2	ICETHK	Tuthill et al.	1998	The model could simulate an equilibrium ice jam profile in conjunction with the HEC-2 backwater model.	-
3	HEC-RAS	U.S. Army	1998	The model could calculate ice regime under different hydraulic conditions, such as open-water flow, sheet-ice cover, ice jam and any combination of above hydraulic conditions.	
4	RICE	Lal, Shen	1991	The concept of two-layer ice transportation, undercover ice accumulation were put forward and simulated respectively.	River ice and ice cover progression simulation on St. Lawrence River in America and Peace River in the Northern British Columbia
5	RICEN	Shen et al.	1995	Super cooling phenomenon and Anchor ice could be simulated. Ice transportation ability equation was used to simulate undercover deposition and erosion instead of critical velocity criterion; The effect of wind, artificial ice-breaking and flow resistance caused by moving ice were included.	Ice jam simulation on Niagara River in 1995
6	DynaRICE	Shen et al.	2000, 2003	2-D dynamic ice transport model, which could be used to simulate dynamic ice jam and surface ice transport.	River ice simulation on Grass Island Pool Area, Niagara Rivers, Mississippi River; the simulation about ice jam time and location on Shokotsu River
7	Dynamics model about surface ice, transportation and ice jam	Lu et al.	1999	Lagrangian discrete method was applied to simulate the surface ice transportation and ice jam.	Ice regime simulation on Niagara Rivers
8	River network ice jam model based on Neural network approach	Massie	2001	This is the first time to use the Neural network approach to forecast the ice jam.	-
9	CRISSP 1D&2D	Shen	2006	The model could simulate the river networks with internal structures and mixed flow conditions and more complicated river system.	Freeze-up ice conditions simulation on the Nelson River; and ice formation simulation on the Red River near Netley Cut and so on

2.3.2 Ice model developments and limitations

The ice model developments and limitations are summarized in Table 2.3, after that a specific introduction about each ice model will be shown.

Table 2.3 Ice model developments and limitations

Model classification	Model name	Main characteristic	Advantages	Limitation
1-D steady state model	HEC-2	Based on the static balance between the external and internal forces to simulate the static ice jam.	Analyze the channel by ice jam thickness and roughness.	Ignore the ice dynamic condition.
	ICETHK		Easily calculate ice-affected stages.	
	HEC-RAS		Calculate under different hydraulic conditions.	
1-D unsteady state model	RICE	A comprehensive 1-D model to simulate the ice flood with unsteady state flow.	The concept of two-layer ice transportation. The assumption of flow with a floating ice cover. The simulation of skim ice and border ice formation. The critical velocity criterion in undercover ice accumulation. The simulation of growth and decay of ice cover.	Lack of detailed consideration to complex flow patterns and river geometry.
	RICEN		River networks simulation. Super cooling phenomenon and anchor ice simulation. Ice transportation ability equation in undercover deposition and erosion Simulation. The effect of wind, artificial ice-breaking and flow resistance.	
	CRISSP1D		River networks with internal structures and mixed flow conditions and more complicated river system. Weather data input method improvement.	
2-D model	DynaRICE	A comprehensive 2-D model to simulate the ice flood with unsteady state flow.	A 2-D dynamic ice transport model, which could be used to simulate dynamic ice jam and surface ice transport.	Lack of detailed consideration on the third dimension.
	CRISSP2D		Complicated flow condition simulation. Super cooling phenomenon simulation. Dynamic surface ice transportation and ice jam dynamic simulation.	

1-D steady state model

The 1-D steady-state models are based on the static ice jam theory, namely that the formulas to determine the final thickness of water surface ice can be deduced according to the static balance of internal and external forces on a floating ice block. Pariset and Hausser (1961) put forward an accumulation theory about water surface ice. They thought that the mode of surface ice accumulation could be divided into juxtaposed cover, thickened narrow jam, and wide-channel jams. According to the static balance of internal and external forces on the floating ice block, they deduced the formulas to determine the final thickness of water surface ice. Based on their work, a lot of researchers continued to refine and extend the accumulation theory, such as Uzuner and Kennedy (1976), Beltaos (1983), Beltaos and Wong (1986), and also created a lot of models based on static ice jam theory to simulate ice jams, such as HEC-2 (U.S. Army, 1990), ICETHK (Tuthill et al., 1998), and HEC-RAS (U.S. Army, 1998). They could be used to simulate the ice regime by the solution of two governing equations, namely ice jam force balance equation and ice jam energy balance equation.

Standard step method is used to solve the two equations. The basic assumption is that the flow is steady, gradually varied, and one-dimensional, and that river channels have small slopes (less than 1:10). The basic equation is the force and energy balance equations; a standard step method is used to solve the equation. However, the limitation of one-dimensional steady-state models is that they ignore ice dynamic conditions.

HEC-2

HEC-2 model is built up to calculate water surface profiles under steady gradually varied flow condition in the natural or man-made channels. The basic assumptions of HEC-2 model are:

- Flow is one-dimensional steady and gradually varied.
- River channels have small slopes (less than 1:10).

The main calculation procedure is to get the solution of one-dimensional energy equation and energy loss equation due to friction evaluated with Manning's equation. The procedure is called standard step method, namely assume the water surface elevation at the one end of reach, and then use the two equations to calculate the water surface elevation at the same end of the reach, then compare them, repeat the procedure until a defined difference is reached. When it comes to river ice cover, the effect of ice cover is taken into consideration by calculating the total Manning's coefficient. The limitation of the model could not be used to solve the problem about the hydraulic stability of broken ice cover.

ICETHK

ICETHK is a kind of software that could be used to simulate an equilibrium ice jam profile in conjunction with the HEC-2 backwater model. The basic assumptions of ICETHK model are:

- One-dimensional steady-state flow.
- Ice is considered stationary.
- The ice jam is assumed as floating ice jam.
- Assume an equilibrium reach of ice jam, where the flow is uniform, ice thickness is constant, and downstream forces acting on the ice cover are resisted entirely by friction at the banks.

The main calculation procedure is that the hydraulic result is calculated by HEC-2 model with an ice cover, and then the result is used to produce new values of ice thickness and ice roughness for the simulated reach by ICETHK model, after that the HEC-2 model recalculate the hydraulic condition with the updated vales of ice thickness and ice roughness from the previous result of ICETHK model. The iteration between the HEC-2 and ICETHK continues until the ice thickness change between the successive iterations is acceptably small. The result of ICETHK is ice thickness and roughness of ice accumulation. The limitation is that the ICETHK model could not address the ice motion or transport problem, and also could not predict where and when an ice jam could occur.

29

HEC-RAS

According to similar differential equations, HEC-RAS was built up. Compared with HEC-2 and ICETHK, the main progress is that it could calculate ice regime under different hydraulic conditions, such as open-water flow, sheet-ice cover, ice jam or any combination of above hydraulic conditions. The two assumptions in the force balance equation: firstly, the longitudinal stress (along stream direction), the accumulation thickness, and the shear stress applied to the underside of the ice by the flowing water are constant along the width; secondly, none of the longitudinal stress is transferred to the channel banks through changes due to stream width, or horizontal bends.

Based on the energy balance equation and force balance equation, ice thickness could be calculated, HEC-RAS model could use two different ways to simulate the ice-covered river: modelling ice cover with known geometry and modelling a wide-river ice jam. In order to start the scheme, the first trial value of ice jam thickness should be provided, and then the energy equation is solved by standard step method from the downstream end. Next, the ice jam force balance equation is solved from the upstream to the downstream, the energy equation and the ice jam force balance equation should be solved alternately until the ice jam thickness and water surface elevations converge to fixed values at each cross-section. The limitation of HEC-RAS model is that it ignores the ice dynamic conditions, because under the natural conditions the flow under the ice cover is almost unsteady (Shen, 2010).

1-D unsteady-state models

Due to the limitation of one-dimensional steady-state models, the development of one-dimensional unsteady-state models, such as RICE (Lal and Shen, 1991), RICEN (Shen et al., 1995), and the Comprehensive River Ice Simulation System (CRISSP1D) model (Chen et al., 2006), was promoted. These models are based on the assumption of unsteady-state flow; the governing equations are one-dimensional Saint Venant equations (i.e. mass and momentum conservation equations with floating ice) and they can be used to simulate the entire ice process in rivers during the winter season. The basic assumptions of the RICE, RICEN, and CRISSP1D models are that they ignore the following:

- The effect on the water body mass balance due to the change in ice phase.
- The river has a floating ice cover.
- Two-layer ice transportation theory.
- Suspension ice is full of the suspension ice layer.
- The thickness of the ice layer on the water surface is equal to the thickness of the ice block floating on the water surface.

However, the limitation of these models is that they lack detailed consideration of complex flow patterns and river geometry (Shen, 2010).

RICE

Due to the limitation of one-dimensional steady-state model, a comprehensive model named RICE model was built to simulate the ice flood with unsteady state flow. The model could be divided into three main components, namely river hydrodynamic model, thermodynamic model and ice dynamic model. What is more, the model could be subdivided as river hydraulics module, water temperature and ice concentration distributions module, ice cover formation module, ice transport and cover progression module, undercover deposition and erosion module, thermal growth and decay of ice covers module, the specific components could be seen in Figure 2.4. The main improvement in the RICE model is that a two-layer ice transportation concept is created, which means the ice discharge consists of surface ice discharge and suspended ice discharge. The specific improvements could be seen as following:

- In the river hydraulics component, the flow with a floating ice cover is described by one-dimensional De Saint Venant equations (mass and momentum conservation equations). The equations are solved by an implicit finite-difference method.
- In the thermal component, the one-dimensional advection-diffusion equation is used to describe the water temperature and depth-averaged frazil ice concentration distributions along the river. The equations are solved by a Lagrangian-Eulerian method.

31

- To simulate the ice transport, a two-layer formulation is used in which the total ice discharge is considered to be made up by the discharge of surface ice and suspended ice. The formation of skim ice and border ice is also included.
- The undercover ice accumulation is formulated according to the critical velocity criterion.
- The growth and decay of ice cover is simulated by finite-difference formulations.

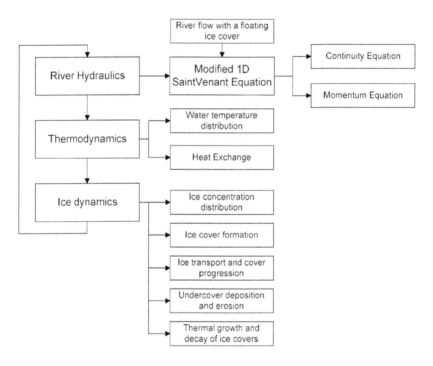

Figure 2.4 Specific components of RICE model

RICEN

Based on the RICE model, the model was updated as RICEN model (Figure 2.5), the main characteristic is that RICEN model could be used to simulate the river network compared with RICE model, and also there are a lot of improvements based on RICE model:

- Super cooling phenomenon and anchor ice could be simulated.

- Ice transportation ability equation is used to simulate undercover deposition and erosion instead of critical velocity criterion (Shen and Wang, 1995).

- The effect of wind, artificial ice-breaking, and flow resistance caused by moving ice is included.

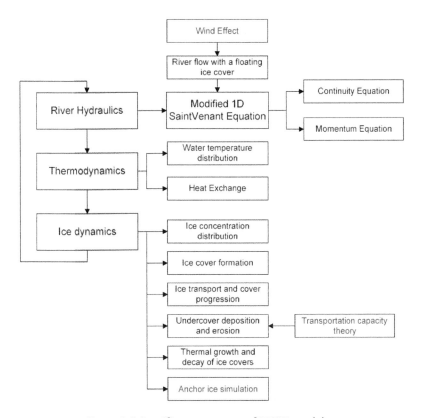

Figure 2.5 Specific components of RICEN model

Because RICE and RICEN models are similar, and RICEN model is based on RICE model, hence they will be introduced together. The basic assumptions are:

- Ignore the effect on the water body mass balance due to ice phase change.

- Assume the river with a floating ice cover.

- Assume two-layer ice transportation theory.

- Assume that suspension ice is full of the suspension ice layer.

- Assume that the ice layer thickness on the water surface is equal to the ice block thickness floating on the water surface.

Main input data: water temperature, air temperature, water level, discharges, and geometric data could be further divided into cross section and bed elevation.

Main output data: the ice front position, ice cover distribution and thickness, thermal ice thickness, frazil ice thickness and concentration, water surface elevation, discharge, and water temperature.

The main calculation procedure of the RICE model is shown in Figure 2.6:

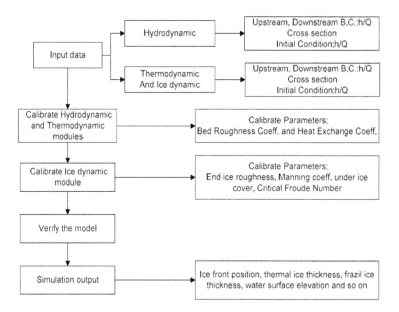

Figure 2.6 RICE model calculation procedure

Model applications: The RICE and RICEN models have been used to simulate the ice regime on several rivers, such as the upper St. Lawrence River near New York and Peace River in northern British Columbia (Li et al., 2002) to the RICE model; Ice jam simulation in 1995 on Niagara River to the RICEN model.

CRISSP1D

Due to the fact that RICE model and RICEN model have been applied on a lot of rivers, in order to improve the model, the latest version of RICE model was created and named as Comprehensive River Ice Simulation System (CRISSP1D) model (Figure 2.7). The main characteristic is that it could be used to simulate the river networks with internal structures, mixed flow conditions, and more complicated river system. Also, there are a lot of improvements based on RICE and RICEN model:

- The simulation of freeze-up ice discharge, under-cover transport, ice cover stability and breakup is improved.
- The weather data from different weather stations could be used in the model and be determined by the distance between the river reach and the nearest weather station.
- Boundary conditions could be more than one with different kinds.

Because CRISSP1D and CRISSP2D are similar, hence, basic assumption, model input data, model output data, calculation procedure, and field application will be introduced together with CRISSP2D. The limitation of the one-dimensional unsteady state models that they are lack of detailed consideration to complex flow patterns and river geometry (Shen, 2010).

Two-dimensional models

Based on the limitations of one-dimensional unsteady-state flow, two-dimensional models have also been developed, such as DynaRICE (Shen et al., 2000) and CRISSP2D (Liu and Shen, 2006) that can be used to simulate the ice regime. These models solve the two-dimensional depth-integrated hydrodynamic equations for shallow water flow. The two basic assumptions are that the movement of the surface ice layer is continuous, and ice is a kind of continuous medium. A finite element method is used to solve the equations. However, the limitation is that they lack detailed consideration of the third dimension.

DynaRICE

DynaRICE is a two-dimensional dynamic ice transport model, which could be used to simulate dynamic ice jam and surface ice transport. In the model, a finite element method is used to solve the hydrodynamics equation, the surface ice transportation and hydrodynamics are under the control of wind force, gravity, flow force, the interaction among ice cubes, and the interaction between ice cubes and river banks; and at the same time, the transportation and hydrodynamics are simulated by a Lagrangian Smoothed Particle Hydrodynamics (SPH) method.

Basic Assumptions:

- Movement of surface ice layer is continuous.
- Ice is a continuous medium.

Model Input data:

- Upstream boundary condition (water level/discharge).
- Downstream boundary condition (water level/discharge/rating curve).
- The initial condition: ice thickness, ice concentration, and water depth/discharge from the whole area.
- Ice supply at the upstream boundary condition (surface ice concentration and thickness).
- Bathymetry data: bed elevation and cross-sections.

Model Output data:

- Distribution of ice particles.
- Distribution of water depth under the ice.
- Distribution of ice thickness.
- Distributions of longitudinal profiles of ice, water surface elevation, and velocity.
- Variations of ice volume and ice velocity with the simulation time.

The main running procedure of DynaRICE model is similar to the CRISSP model, which will be discussed in the section of CRISSP model.

Field application:

DynaRICE is a two-dimensional river ice dynamics model, and it has also been applied on several rivers to understand the ice jam evolution, ice booms design, and navigation structures design and so on, such as the application on the Niagara Power Project to study the ice control and ice-period operation, the application on the Missouri-Mississippi River to study ice jam formation, and the application on the Shokotsu River in Hokkaido of Japan to study breakup jam (Shen, 2000).

CRISSPD2D

CRISSP2D model is based on the CRISSP1D and DynaRICE, and it becomes more powerful compared with CRISSP1D and DynaRICE, and also there are a lot of improvements based on CRISSP1D and DynaRICE model:

- Complicated flow conditions (such as super-critical, sub-critical flow and dry-wet bed conditions) can be simulated.
- Super cooling phenomenon could be simulated in the water temperature distribution module.
- Frazil ice, border ice and surface ice runs could be included in the process of freeze-up.
- Dynamic surface ice transportation and ice jam dynamic are included in the model.

DynaRICE and CRISSP2D model could be used to simulate the ice regime by the solution of the governing equations, namely two-dimensional depth-integrated hydrodynamic equations for shallow water flow. A finite element method is used to solve the equations.

Basic Assumptions:

Because CRISSP1D and CRISSP2D are based on RICE and DynaRICE models, hence, the basic assumptions are the same for RICE and DynaRICE models.

The main running procedure of CRISSP1D & CRISSP2D model are indicated in Figure 2.6.

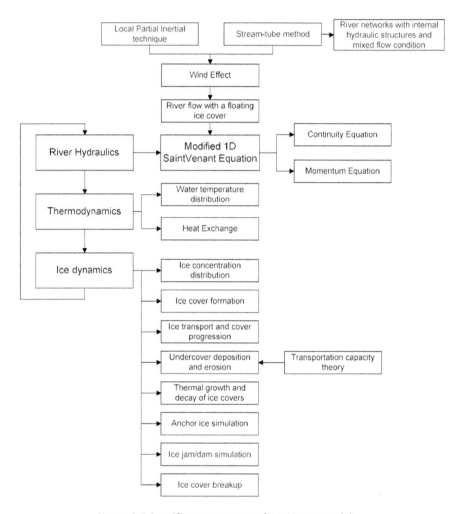

Figure 2.7 Specific components of CRISSP1D model

Field application:

CRISSP1D and CRISSP2D have been applied to several rivers and lakes to simulate the ice regime. For example, CRISSP2D has been applied on the Nelson River to simulate freeze-up ice conditions (Malenchak et al., 2008); and also applied on the Nelson River to model ice formation on the Red River near Netley Cut (Haresign and Clark, 2011). Apart from that, CRISSP2D could be also used to test the potential for anchor ice growth after the Conawapa Power Generation Station was constructed and operational (Morris et al., 2008).

The limitation of the two-dimensional models is that there is a lack of detailed consideration on the third (vertical) dimension.

Data driven models

The potential for using fuzzy expert systems to forecast the potential risk of ice jam was discussed by Mahabir et al. (2002). Based on this research, more research has been carried out (Mahabir et al., 2006; Wang et al., 2008). Data-driven modelling has been applied to the ice regime simulation several times. However, the limitation of using such methods is that ice processes cannot be understood by data driven models.

2.4 Critical assessment of existing ice flood models

The ice flood modelling is more complex comparing with the flood modelling triggered by rainstorm, which means the building of the ice flood modelling should take into account the thermal factor and coupling of hydrodynamics with thermodynamics as indicated as Figure 2.1. After the ice frazil is formed in the water, the river water flow transforms into two phase flow, and ice frazil and water exchange according to the heat loss or gain of water body. And the movement and distribution of ice frazil is different with the that of sediment in the water body, due to the density of ice is lower than that of water, thus the ice frazil is moving upward and sediment is moving downward. Therefore, the simulation of ice frazil development is complex and very important for ice flood modelling.

For the one-dimensional ice flood modelling, as previous sections mentioned, due to the limitation of one-dimensional steady-state models, the development of one-dimensional unsteady-state models, such as RICE, RICEN, and CRISSP1D, was promoted. These models are based on the assumption of unsteady-state flow; the governing equations are one-dimensional Saint Venant equations (i.e. mass and momentum conservation equations with floating ice) and they can be used to simulate the entire ice process in rivers during the winter season. Shen (2010) thought these models are lack detailed consideration of complex flow patterns and river geometry. The basic assumptions of these models are that they ignore the following: (i) the effect on the water body mass balance due to the change in ice phase; (ii) the river has a floating ice cover; (iii) two-layer ice transportation theory; (iv) suspension ice is full of the suspension ice layer; (v) the thickness of the ice layer on the water surface is equal to the thickness of the ice block floating on the water surface. Therefore, these models still have gaps need improvement.

For the aspect of the effect on the water body mass balance, due to the change in ice phase, which means these models are lack consideration of two phases exchange, and should consider water balance to reflect the actual behaviors. Especially for the Ning-Meng reach of the Yellow River basin, which the river channel is flat and wide, during the winter freeze-up period, with the water level increase, a large amount of water go into the floodplain and become the channel water storage and ice cover there, which cause the water loss and unbalance, and during the breakup period of the river, the channel water storage releases from the floodplain and the melting ice back into the channel to result in more water volume. Therefore, for the Ning-Meng reach, the channel water storage should be considered in the mathematical representation of the ice flood modelling.

During the river freeze-up period, the ice cover roughness plays an important role to influence the river flow, thus its influence could not be neglected in simulation of the river ice flood modelling. For the aspect of the river has a floating ice cover, which means these models take the roughness of ice cover as the end roughness of ice cover and decay parameter as constant, and lack consideration of ice cover roughness varies during the ice cover formation and decaying procedures, and decay parameter should varies according to different ice regime conditions. Therefore, the simulation of these river ice flood models could not reflect the real behaviors of river flow with ice covered. It requires using the in situ observation data and actual ice regime conditions to quantify the ice cover roughness and decay parameter at different hydrometeorological conditions in order to reflect the actual situations.

In case of two-layer ice transport theory, these models use a two-layer model for surface ice and river flow beneath. The thickness of ice layer on top of the water surface is equal to the thickness of an ice block floating on the water surface, which means most of the ice cover is submerged. This can be dealt with by introducing a coefficient of submergence in the mathematical representation of ice flood modelling.

For these models, the critical value of discharge for which the river breakup appears is a constant value that has to be determined before using the model, by experiments on site. The critical discharge value should be varied with the parameters such as ice cover thickness and air temperature and channel geometry etc. for actual conditions. Furthermore, these models are lacking a river freeze-up consideration is influenced by discharge and air temperature and channel geometry. In order to overcome these shortages, it is necessary to setup river freeze-up and breakup empirical criteria based on the ice regime conditions to reflect the actual situation.

Chapter 3 Field Campaign and Data Analysis

3.1 Study area

The Yellow River is the second longest river in China after the Yangtze River, and the sixth longest in the world. Originating on the Qinghai-Tibetan plateau in western China, it flows across Qinghai, Sichuan, Gansu, Ningxia, Inner Mongolia, Shaanxi, Shanxi, Henan and Shandong Provinces or Autonomous before draining into the Bo Hai Sea, extending across 95°53'-119°05' East, 32°10'-41°50' North, about 1,900 km east-west, and 1,100 km south-north, with a total watershed area of 794,712 km^2 (including inner watershed area of 42,269 km^2), a total length 5,464 km and surface fall of 4,480 m, under the influence of cold air from Siberia and Mongolia in winter period. As the cradle of the Chinese civilizations and the centre of China's current political, economic and social development, the river is known as 'the mother river of China'.

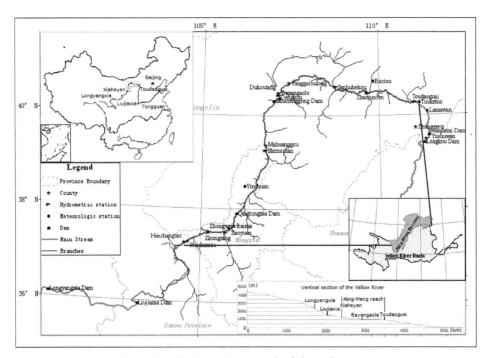

Figure 3.1 The Ning-Meng reach of the Yellow River

Table 3.1 Channel characteristics of the Ning-Meng reach

Autonomous region	Section	Channel type	Channel length (km)	Average channel width (m)	Main channel width (m)	Channel slope (‰)	Bend ratio
Ningxia	Nanchangtan-Zaoyuan	valley type	135	200-300		0.8-1.0	
	Zaoyuan-Mahuanggou	transition type	262	500-1000		0.1-0.2	
Inner Mongolia	Mahuanggou-Wuda bridge	valley type	69.0	400	400	0.56	1.5
	Wuda bridge-Sanshenggong	transition type	106.6	1800	600	0.15	1.31
	Sanshenggong-Sanhuhekou	wandering type	205.6	3500	750	0.17	1.28
	Sanhuhekou-Zhaojunfen	transition type	126.2	4000	710	0.12	1.45
	Zhaojunfen-Lamawan	bend type	214.1	3000 in upstream; 2000 in downstream	600	0.10	1.42
	Lamawan-Yushuwan	valley type	118.5				
Total	Nanchangtan-Yushuwan		1237				

The Ning-Meng reach belongs to temperate continental monsoon climate. In winter, which is controlled by the strong Mongolia cold high pressure, and with the effect of high northwest air flow, the weather is cold and dry, the mean air temperature during the whole winder is below freezing point, and this lasts for 4 to 5 months. And this reach is the lowest air temperature place in winter besides the source area in the Yellow River basin.

The Ningxia reach starts form Nanchangtan, flowing through Heishanxia and Qingtongxia, and ends at Mahuanggou of Shizuishan city, with a total length of 397 km. The reach from Nanchangtan to Zaoyuan, flowing from southwest to northeast, length of 135 km, is narrow-valley type section with steep slope which the channel slope is 0.8‰-1.0‰ and quick water flow, which belongs to uncommon freezing reach, and just freezes in very cold year; the reach from Zaoyuan to Mahuanggou, flowing from south to north, length of 262 km, with gentle slope which the channel slope is 0.1‰-0.2‰ and slow water and low temperature, belongs to common freezing reach. After the operation of Qingtongxia and Liujiaxia reservoir, for the discharge and the water temperature increased, the uncommon freezing reach extends downstream to Baima of Zhongning, and the reach downstream of Qingtongxia reservoir 40-90 km also becomes the uncommon freezing section. The Inner Mongolia reach lies in the top north of the Yellow River. Starting from Mahuanggou of Shizuishan city, it flows from south to north, flowing through Sanshenggong hydro-junction, then turns to northeast direction, flowing to Yanggaibulong, then turns to southeast direction, flowing to Sanhuhekou, then turns to flow from west to east, flowing to Baotou, then turns to southeast direction, flowing to Lamawan, then turns to flow from north to south, flowing through Wanjiazhai reservoir, and finally ends at Yushuwan, with a total length of 840 km. Generally, the Inner Mongolia reach exhibits the characteristics of large top width and gentle channel slope, many bends and large bend ratio, the channel slope is large in the upper stream and small in the lower stream, the river bed is deep and narrow in the upper stream and turns to shallow and wide in the lower stream, which the maximum width is 400-1,200 m at Dukoutang and the minimum one is 200-400 m at Zhaojunfen, and the channel slope is just 0.09 in the section between Zhaojunfen and Toudaoguai, which presents the river bed characteristics in plain. From Bayangaole to Toudaoguai section, there are 69 river bends and the maximum river bending is 3.64. Before the operation of Wanjiazhai reservoir, the event of freezing rarely occurs in the section between Toudaoguai and Wanjiazhai, where mainly occur ice run, with just bank ice and frazil ice run, for steep channel slope and rapid flow velocity. While after the operation of Wanjiazhai reservoir, it is easy to occur ice accumulation and ice jam, for the reduction of channel slope and water velocity in the reservoir region and the end of back water region, as well as the reduction of ice discharge capacity, and the ice accumulation usually extends upstream, as a result of the original uncommon freezing section comes to common freezing section.

For the sedimentation, as the reach of Bayangaole to Toudaoguai, there is 0.28 billion ton sediment erosion from 1960 to 1986 and annual mean sediment erosion of about 0.01 billion ton, but since 1986 which the Longyangxia reservoir put into operation, there is total

1.38 billion ton sediment deposition from 1986 to 2010 and annual mean sediment deposition is about 0.06 billion ton, especially for the Shidakongdui reach the river bed increase 1.5 to 2 m. Therefore, in recent 20 years, the channel sediment deposition increased from Bayangaole to Toudaoguai section, and the main channel shrinkage which decreases the channel capacity and the flood-carrying capacity, especially in ice flood period, the possibility of ice jams increase and the bank-full discharge reduced from 4,000-5,000 m^3/s in 1970s and 1980s to present 1,000-1,500 m^3/s (Huo et al., 2007). With the river bed deposition can reduce ice transport capacity and the flow capacity under ice cover further. Most of the water stops in floodplains and channel water storage increases. During the breakup, the ice flood peak and flow dynamics get small due to the storage retained in floodplains and releases slowly. The ice flood often shows twin peaks or a low peak generally.

3.2 Data availability

The datasets of the Ning-Meng reach (Table 3.2) consist of three parts, first part is historic conventional observed data, such as meteorological and hydrological data from meteorology and hydrometric stations; second part is new added observed meteorology and ice regime data at selected hydrometric stations and cross-sections for parameters calibration and verification of the ice flood modelling, the last part is the cross-section and DEM data for river channel construction of the ice flood modelling.

The purpose of the historic data is to be used for data analysis and some parameters determination of the ice flood modelling. The meteorological stations include Yinchuan, Linhe, Baotou and Tuoketuo, and the time series from 1954 to 2010. The observed item is daily air temperature. The hydrological stations are Lanzhou, Shizuishan, Bayangaole, Sanhuhekou and Toudaoguai, the observed items including daily water temperature, ice cover thickness, discharge, water level, date for ice run and freeze-up and breakup, ice flood peak and occurrence time, and the time series from 1950 to 2010, and using the discharge data of different cross-section and considering the river channel routing time to calculate the channel water storage data between the calculated cross-sections during the freeze-up and breakup period.

Table 3.2 Datasets of the Ning-Meng reach

Classification	Item	Station	Period
Meteorological data	Air temperature.	Yinchuan, Linhe, Baotou, Tuoketuo	1954-2010; daily.
Hydrology and ice regime data	Discharge, water level, water temperature, ice cover thickness, date for ice run and freeze-up and breakup, ice flood peak.	Lanzhou, Shizuishan, Bayangaole, Sanhuhekou and Toudaoguai	1950-2010; daily or yearly.
Automatic meteorology stations	Air temperature, humidity, wind speed and direction, air pressure, total radiation.	Sanhuhekou, Baotou, and Toudaoguai	Winter of 2013/2014 and 2014/2015; 1 hour.
Added hydrology and ice regime data	Initial ice run and freeze-up position and time, ice cover thickness, flow velocity and water level under the ice cover, water temperature, ice concentration, ice frazil velocity and size.	Sanhuhekou, Baotou, and Toudaoguai	Winter of 2013/2014 and 2014/2015.
River ice survey data	Initial ice run and freeze-up position and time, ice cover thickness etc.	From Sanhuhekou to Toudaoguai	Winter of 2013/2014 and 2014/2015.
Cross-section data	DEM and river bed load material.	For Ningxia reach 69 cs, for Inner Mongolia reach 167 cs.	For Ningxia reach in 2009 after flood season, for Inner Mongolia reach in 2012 after flood season.

3.3 Field campaign

Ice regime information observation and forecasting is the most important scientific supporting measures for ice flood control, reservoir regulation and decision making. Tracking ice formation from observations and combining them with numerical model predictions for advanced warning requires proper understanding of all scientific issues that play a role. However, it is not possible to make specific predictions because our physical understanding remains incomplete, thus the main challenge is how to accelerate the pace of discovery and bridge the major knowledge gaps. In the case of the Yellow River, ice floods impose a threat every year, which is why the YRCC is putting considerable effort in verifying theoretical formulations with actual field measurements in order to better understand the scientific mechanisms that play a role.

The new added observed meteorology data including 3 automatic meteorological observation stations at Sanhuhekou, Baotou, and Toudaoguai hydrometric stations (Figure 3.2), the observed items are air temperature, humidity, wind speed and direction, air pressure, and total radiation, and time interval is 1 hour, and observed period is the winter of 2013/2014 and 2014/2015.

Figure 3.2 Automatic meteorology stations

New added observed ice regime data including initial ice run and freeze-up position and time, ice cover thickness, flow velocity and water level under the ice cover, water temperature, ice concentration, ice frazil velocity and size etc. at Sanhuhekou, Baotou, and Toudaoguai hydrometric stations, and for Toudaoguai install H-ADCP to realize the under ice cover flow velocity continuous observation. And survey ice regime data from Sanhuhekou to Toudaoguai, and observed period is the winter of 2013/2014 and 2014/2015. Figure 3.3 shows some instruments used for ice regime observation and river ice survey.

Figure 3.3 Some instruments used for ice regime observation and river ice survey

Cross-section data including DEM data and river bed load material for Ningxia part and Inner Mongolia part, total 236 cross-sections, on average 4 km interval for each cross-section, which Ningxia part has 69 cross-sections, Inner Mongolia part has 167 cross-sections. In this research, according to the data availability, the ice regime characteristics to select the reach from Sanhuhekou to Doudaoguai were used to setup the ice flood model. The cross-section sketch map from Sanhuhekou to Toudaoguai part is shown in Figure 3.4. Figure 3.5 shows the variation of channel erosion and deposition at Toudaoguai hydrometric station from 2012 to 2013.

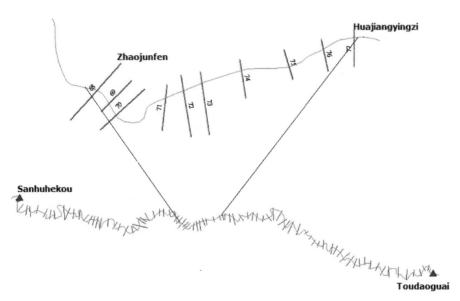

Figure 3.4 Cross-section sketches from Sanhuhekou to Toudaoguai

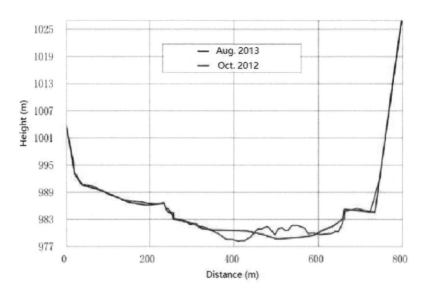

Figure 3.5 Cross-section sketch for Toudaoguai hydrometric station

3.4 Data analysis of longer term trends

Ice regime is the result of comprehensive influences of temperature, water situation, channel condition and human activities. The ice regime and human activity have impacted each other in the Ning-Meng reach. The upstream reservoirs of Liujiaxia and Longyangxia (see Figure 3.1), which were put into operation in 1968 and 1986 respectively, provided very important regulations for the ice regime of the Ning-Meng reach. The reservoir operation can influence the ice regime downstream by changing the discharge and water temperature. Some projects such as bridges and pontoons in channel can cause resistance to ice and water flows, and also influence the ice regime. On the other hand, the existence of ice can impact the safety and benefit of the water conservancy projects. Due to the effect of natural factors such as river course conditions, hydrological and meteorological conditions and human activities, the ice regime of the Ning-Meng reach is different every year. This is due to early or late date on which river ice run, freeze-up and breakup, long or short of the stable freeze-up duration, thick or thin ice cover thickness, possible ice clog and jam disasters, channel water storage increment during the freeze-up and ice-covered period, channel water storage release amount and the ice flood peak. With the development of society and economy, the ice regime characteristics have changed. At the same time, ice characteristics have an impact to local industry, agriculture and human life. Lately, the ice disasters, in particular the ones due to freeze-up or breakup, attracts more and more the attention of the water authorities and local government. It is important to know the ice regime characteristics for ice forecasting and ice flood prevention and alleviation.

3.4.1 Moment of river ice run and freeze-up and breakup

In winter when the Ning-Meng reach of the Yellow River freezes up, the first freeze-up takes place from Sanhuhekou to Toudaoguai sections of the river. According to the statistical analysis of the river ice run, freeze-up and breakup date of the Ning-Meng reach (Table 3.3), the first freeze-up moment is in the period from the middle of November to December, every year. Since 1970, the earliest freeze-up date is on November 13, which occurred in 1976, and the last one is on December 30, which occurred in 1989. During the river breakup period, the latest breakup date is in the middle or end of March, while the earliest breakup date is on March 12 which occurred in the ice flood period of 1997/1998. The latest breakup date is on April 4 which occurred in the ice flood period of 1975/1976. From Table 3.3 it can be seen that in the last 20 years, comparing with the data of 1970-1990, and the river ice run and freeze-up date is happening 1 to 2 days later, while the breakup is 2 days earlier. Due to the nature of phenomenon of the ice formation, a duration of 1-2 days is considered adequate. In the last 10 years, the river freeze-up dates are 2 to 4 days later, while breakup dates are 2 days earlier.

Table 3.3 Moment of river ice run, freeze-up and breakup of the Ning-Meng reach

Period		Ice run date (month-day) (Year)	Freeze-up date (month-day) (Year)	Breakup date (month-day) (Year)
1950-1970	Mean	11-16	11-30	3-9
	Earliest	11-4 1969	11-7 1969	2-28 1966
	Latest	11-23 1954,1964,1965	12-14 1961	3-18 1970
1970-1990	Mean	11-19	12-2	3-26
	Earliest	11-8 1981	11-13 1976	3-15 1989
	Latest	11-26 1980	12-30 1989	4-4 1975
1990-2010	Mean	11-21	12-3	3-24
	Earliest	11-8 2000	11-16 2000	3-12 1997
	Latest	11-30 2006	12-16 1992	3-31 1984,2009
1990-2000	Mean	11-17	12-2	3-23
	Earliest	11-8 2000	11-16 2000	3-12 1997
	Latest	11-27 1994	12-16 1992	3-30 1995
2000-2010	Mean	11-23	12-4	3-24
	Earliest	11-8 2000	11-16 2000	3-14 2001
	Latest	11-30 2006	12-11 2007	3-30 2004

3.4.2 Stable freeze-up duration

Stable freeze-up means the river formed the stable ice cover from bank to bank, or the moving part less than 20 percent of the total river course area, and can be classified as thermal freeze-up and packed freeze-up due to the river frozen conditions. Table 3.4 shows the statistical characteristics values of the stable freeze-up duration on the Ning-Meng reach at different hydrometric stations. From Table 3.4, it can be seen that the mean annual freeze-up duration is about 117 days in the Ning-Meng reach, and the longest one is 150 days, which occurred in the winter of 1969/1970. The shortest period of freeze-up is of 76 days, which occurred in the winter of 1989/1990. The mean annual freeze-up duration of Shizuishan station is of 60 days which is the shortest one in the Ning-Meng reach, and at Sanhuhekou is the longest freeze-up period of 108 days. Comparing the periods from 2000 to 2010, with that of 1950-2010, the stable freeze-up duration of the Ning-Meng reach is shorter by 3 days, Shizuishan 22 days shorter, Bayangaole 17 days shorter, Sanhuhekou and Toudaoguai 6 days shorter respectively.

Table 3.4 Stable freeze-up duration of the Ning-Meng reach and hydrometric stations

Item		Shizuishan	Bayangaole	Sanhuhekou	Toudaoguai	Ning-Meng reach
Mean of 1990-2010 (days)		39	78	100	93	112
Mean of 1950-2010 (days)		60	93	108	100	117
Mean of 1990-1999 (days)		33	78	102	95	111
Mean of 2000-2010 (days)		38	76	102	94	114
Shortest	Days	12	49	72	53	76
	Occurrence year	2006/2007	1989/1990	1998/1999	2001/2002	1989/1990
Longest	Days	102	124	150	135	150
	Occurrence year	1969/1970	1970/1971 1976/1977	1969/1970	1969/1970	1969/1970

3.4.3 Water temperature

The river ice run is the result of low air temperature. One of the key factors to illustrate the river thermal conditions is water temperature. The heat exchange of atmosphere and river water has an effect on the increase and decrease of the water temperature. In winter when the air temperature is below freezing point, the negative air temperature triggers the water body to lose heat and cool, hence generates ice. In the ice flood period, the operation of the reservoir, do not only adjust the discharges on the downstream, but also the releases have higher water temperature and as such increases the water temperature of the downstream river reach, impacting the flow thermal condition. Table 3.5 shows the water temperature comparison for the month in November and March (the river freeze-up and breakup time) at the 4 hydrometric stations in the Ning-Meng reach. From Table 3.5, it can be seen that after 1968, when Liujiaxia reservoir was put into operation, the water temperature increases 0.7-0.9 centigrade from Shuizuishan to Bayangaole during the month of November of every year, while the water temperature, on the section between Sanhuhekou to Toudaoguai, do not vary so much. The water temperature increases 0.5 centigrade at Shuizuishan in March, while at the other stations it does not vary so much. After 1986, when Longyangxia reservoir was put into operation, it can be noticed that water temperature increases by about 0.7 centigrade from Shuizuishan to Bayangaole during November as compared with the period when only Liujiaxia reservoir was operated. At Sanhuhekou station the water temperature increases by 0.3 centigrade, and at Toudaoguai it decreases by 0.5 centigrade. In March, the water temperature increases by 2.4 centigrade, 1.9 centigrade, 0.7 centigrade and 0.4 centigrade for Shuizuishan, Bayangaole, Sanhuhekou and Toudaoguai respectively. The main conclusion is that after Liujiaxia and Longyangxia reservoirs were put into operation, the discharge operated through the reservoir significantly affect the water temperature at Bayangaole during the ice run and freeze-up period. In the same time on the reach from Sanhuhekou to Toudaoguai there is no evident effect of the reservoir operation, which is to be expected, because of the far away location of this reach from the reservoirs. The reservoir releases significantly affects, as well, the water temperature of the Ning-Meng reach during the breakup period in March.

Table 3.5 Water temperature on Nov. and March at 4 hydrometric stations (Unit: ℃)

Period		Shizuishan		Bayangaole		Sanhuhekou		Toudaoguai	
		Nov.	March	Nov.	March	Nov.	March	Nov.	March
1950-1967	First ten-day	5.6	0.3	4.7	0.1	4.3	0.2	3.9	0
	Middle ten-day	3.1	2.0	2.3	0.6	1.9	0.3	1.4	0.1
	Last ten-day	1.2	5.1	0.8	4.3	0.5	2.2	0.4	1.0
	Monthly	3.3	2.5	2.6	1.7	2.2	0.9	1.9	0.4
1968-1985	First ten-day	7.0	0.7	6.5	0	5.0	0	4.5	0
	Middle ten-day	3.4	2.8	2.8	1.0	1.5	0	1.3	0
	Last ten-day	1.7	5.6	1.2	4.7	0.3	1.7	0.2	0.9
	Monthly	4.0	3.0	3.5	1.9	2.3	0.6	2.0	0.3
1986-1999	First ten-day	6.8	3.1	6.5	1.1	4.7	0	2.9	0
	Middle ten-day	4.5	5.4	4.2	3.5	2.4	0.3	1.2	0.1
	Last ten-day	2.9	7.5	2.6	6.5	0.9	3.6	0.4	2.0
	Monthly	4.7	5.4	4.3	3.8	2.6	1.3	1.5	0.7
2000-2010	First ten-day	7.2	3.2	7.0	1.3	5.3	0.1	4.7	0
	Middle ten-day	4.6	5.4	4.2	3.9	2.8	0.8	2.1	0.8
	Last ten-day	2.8	8.5	2.3	7.4	0.9	4.8	0.6	3.7
	Monthly	4.9	5.7	4.5	4.2	3.0	1.9	2.5	1.6

In the last ten years, from 2000 to 2010, due to the effect of global warming, the winter air temperature increases for the whole Yellow River basin, and especially for the Ning-Meng reach. This could be concluded due to the notably changes of the water temperature for the ice run and freeze-up period during November every year. And the breakup period during March for the Ning-Meng reach. Comparing the water temperature with that of 1986-1994, it is noticeable of an increase of 0.4-1.8 centigrade in the first ten-day of November. And increase of 0.9-1.7 centigrade in the last ten-day of March respectively. The river reach, from Shizuishan to Toudaoguai the affection become evident gradually, and for the Toudaoguai the affection is most evident, and for Shizuishan, Bayangaole, Sanhuhekou and

Toudaoguai in the first ten-day of November increase 0.4 centigrade, 0.5 centigrade, 0.6 centigrade and 1.8 centigrade respectively, in November increase 0.2 centigrade, 0.2 centigrade, 0.4 centigrade and 1.0 centigrade respectively, in the last ten-day of March increase 1.0 centigrade, 0.9 centigrade, 1.2 centigrade and 1.7 centigrade respectively, in March increase 0.3 centigrade, 0.4 centigrade, 0.6 centigrade and 0.9 centigrade respectively.

Therefore both the air temperature variation and reservoirs operation affect the winter water temperature of the Ning-Meng reach, and also influence the freeze-up ice amount, freeze-up length, dates for freeze-up and breakup, and breakup conditions.

3.4.4 Ice cover thickness

After the freeze-up of the Ning-Meng reach, in winter time, the ice cover thickness varies and in general is larger than 50 cm. Table 3.6 shows the maximum ice cover thickness at 4 hydrometric stations on the Ning-Meng reach, the mean maximum ice cover thickness is 81 cm in the Inner Mongolia reach and the maximum one is 109 cm which occur at Bayangaole in 1976. The minimum ice cover thickness is 35 cm and which occur at Toudaoguai in 1977. Since 1986 the air temperature of the Ning-Meng reach has increased in winter, which with the effect of climate change, also clearly reflected the variation of ice cover thickness.

From 1990 to 1999, the mean ice cover thickness is only 55-58 cm, while in recent ten years from 2000 to 2010, it is 53-63 cm. Compare the last 10 years with the ones before 1990, the mean ice cover thickness is thinner by 10 cm. However at some reaches occur the lowest value based on observation, such as at Bayangaole, the maximum ice cover thickness is only 40 cm in the winter of 1994/1995 and 1998/1999, and at Sanhuhekou, the maximum ice cover thickness is also only 40 cm in the winter of 2001/2002.

Table 3.6 The maximum ice cover thickness of hydrometric stations (Unit: cm)

Item	Shizuishan	Bayangaole	Sanhuhekou	Toudaoguai
Mean of 1970-1989	41	81	72	64
Mean of 1990-1999	39	55	56	58
Mean of 2000-2009	36	57	53	63
Maximum ice cover thickness	55	109	95	80
Occurrence year	1982/1983	1976/1977	1979/1980	1982/1983
Minimum ice cover thickness	20	40	40	35
Occurrence year	1978/1979	1994/1995 1998/1999	2001/2002	1977/1978

3.4.5 Channel water storage

When the river freeze-up, the water flow resistance is growth due to the ice cover and ice jam, which result in the flow capacity decrease below the ice cover, and amount of water retained upstream as backwater to rise the water level, some of the water go to the floodplain and formed the channel water storage, Figure 3.6 shows the sketch of the formation of channel water storage. Channel water storage is the water retained in the river channel due to factors such as frozen ice cover during the ice flood period. The channel water storage is different every year with the different air temperature, reservoir discharge and ice regime characteristics every ice flood period. The channel water storage increase as with the ice cover increase after the river freeze-up stably, and when freeze-up length is the longest then the channel water storage also reach its maximum. In that case the channel water storage does not vary so much because the ice cover is stable, when the air temperature starts to increase the ice cover begins to melt and the channel water storage decrease gradually. The water amount are released in the downstream, until the breakup period, hence the channel water storage concentrate to release and generate the ice flood peak.

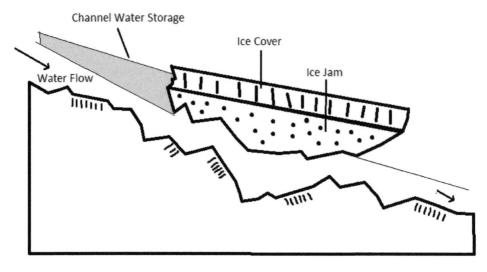

Figure 3.6 Channel water storage formation

Figure 3.7 shows the channel water storage variations over the years from 1970 to 2010 during the ice flood period of the Ning-Meng reach, the mean channel water storage is about 1.24 billion m³ from 1970 to 2010, and the maximum one is 1.91 billion m³ which occurred in the ice flood period of 2004/2005, and the minimum one is only 0.45 billion m³ which occurred in the winter of 1996/1997. For the total tendency since 1990, besides the winter of 1996/1997, the channel water storage obviously increasing which comparing with that before 1990. Before 1990, the maximum channel water storage is only 1.4 billion m³, which occurred in the winter of 1976/1977, and before that, the channel water storage maintain at about 1.0 billion m³, and especially from 1982 to 1987, the channel water storage is below 1.0 billion m³. Therefore, the total tendency of the channel water storage increased obviously since 1990. And since 2000, the annual mean of the channel water storage has increased from 1.24 billion m³ to 1.59 billion m³.

After 1990, the channel water storage capacity of Ning-Meng reach increased significantly due to the serious channel sedimentation and shrivel. As a consequence of sedimentation the main channel discharge capacity decrease to a bank-full discharge value below 1,500 m³/s. When freeze-up takes place in winter the ratio of the floodplain water storage and channel water storage increases and as a result the channel water storage increases.

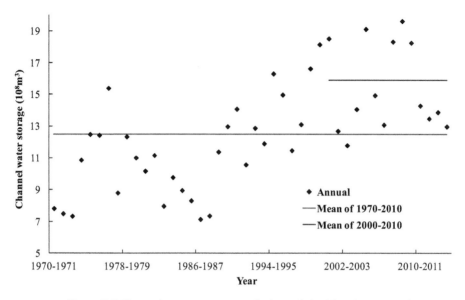

Figure 3.7 Channel water storage variations of the Ning-Meng reach

3.4.6 Water level

Figure 3.8 and 3.9 shows water level variations during the freeze-up period and the breakup period at Sanhuhekou and Toudaoguai stations since 1986. In the last years, due to the serious channel sedimentation and shrivel, the river bed increased, and the main channel discharge capacity decreased, which illustrate the water level of Sanhuhekou and Toudaoguai stations increased during the freeze-up period, especially for the Sanhuhekou, and the water level increase 2 to 3 m.

Analyzing the data shows that water level has the same increasing tendency for the breakup period, which for the Sanhuhekou, the historical first high water level occurred on March 20, 2008 was 1021.22 m and the historical second high water level occurred on March 19, 2009 was 1020.98 m.

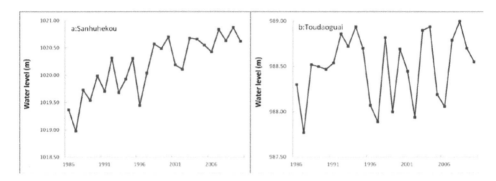

Figure 3.8 Water level variations at Sanhuhekou (a) and Toudaoguai (b) during ice freeze-up

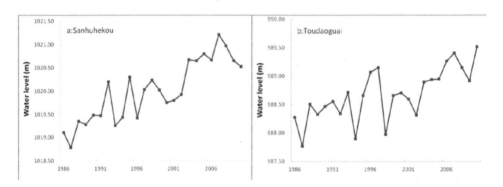

Figure 3.9 Water level variations at Sanhuhekou (a) and Toudaoguai (b) during ice breakup

3.4.7 Ice flood peak and volume

Because of the influence of the geographic location and river flow directions, during the breakup period of the Ning-Meng reach, the melting starts from upstream to downstream, hence the channel water storage releases increase gradually along the river course, and the ice flood peak increase gradually along the river course. The ice flood mainly occurred at the reach from Bayangaole to Toudaoguai, and the maximum ice flood peak usually happened at Toudaoguai, the historical maximum ice flood peak is 3,500 m^3/s in 1968 and the minimum one is 1,000 m^3/s in 1958 and the mean ice flood peak is 2,192 m^3/s from 1950 to 2010. The longest duration of the ice flood is 10 days and the shortest one is about 7 or 8 days. So taking the 10 days water amount as the ice flood volume, the mean one of the Toudaoguai is about 0.95 billion m^3 from 1950 to 2010, and the maximum one is about 1.52 billion m^3 in the winter of 2000/2001 and the minimum one is only about 0.52 billion m^3 in the winter of 1958/1959.

Table 3.7 shows the mean ice flood peak and 10-day water amount of Toudaoguai, after Liujiaxia and Longyangxia reservoir put into operation, the ice flood peak and 10-day water amount increased gradually comparing with the natural conditions, and in the last years after Wanjiazhai reservoir put into operation in 1999, the ice flood peak increase but the 10-day water amount decreases.

Table 3.7 Mean ice flood peak and 10-day water amount of Toudaoguai

Period	10-day flood volume (billion m^3)	Ice flood peak (m^3/s)	
1950-2010	0.95	2192	
1950-1968	0.74	1965	Before Liujiaxia reservoir put into operation
1969-1986	0.92	2441	Before Longyangxia reservoir put into operation
1987-1999	1.06	2355	Before Wanjiazhai reservoir put into operation
2000-2010	1.17	1944	

3.4.8 Ice regime influence factors

Characters of ice regime are the results of comprehensive influences by thermal factors, dynamical factors, channel course conditions and human activities. The thermal factors include solar radiation and scattering radiation; air temperature; water temperature; solar radiation and surface reflection; are important to determine the air temperature, which in turn influences water temperature and ice conditions. Hence the air temperature is the key illustration of thermal factors which influence the ice regimes.

The dynamical factors include discharge, water level, and flow velocity, because discharge influences flow velocity and water level variation. Flow velocity influences the ice formation, as well as ice transport and ice clog etc. Water level variation has close relation with river freeze-up and breakup conditions. If water level is constant it generates thermal breakup, and if water level increases rapidly it will generate mechanical breakup. The variations of flow velocity and water level depend on the variation of discharge, so the discharge is the key illustration of a dynamical factor, which influence the ice regimes evolution.

Channel course conditions include river course location, directions and river cross-sections and boundary characteristics. When the air temperature and discharge do not vary so much, it is very easy to have ice run clog, ice accumulation, freeze-up and ice jams at the local reach, as river narrows reach, meandering and shallow area.

Human activities include the reservoirs regulation, and structures along the river, reservoirs regulation will not only change the channel discharge allocation, but also increase the water temperature, so the influence of reservoir is a dynamical and thermal factor. The structures along the river such as bridges, pontoons, and construction piles and cofferdams influence the river flow directions and boundary conditions.

Air temperature variation

The winter air temperature variations at meteorological stations are shown in Figure 3.10 and the statistical characteristic value of the winter accumulated air temperature is shown in Table 3.8 (Wang et al., 2012). Through analysis of the annual variation of the average winter air temperature and winter accumulated air temperature of the meteorological stations the following can be concluded:

- The winter air temperature changing trends of the meteorological stations are very similar; the variation of the peak and valley point is basically identical.
- The air temperature of the stations of Ning-Meng reach is below freezing point during the winter which lasting 5 months.
- The winter air temperature of the Ning-Meng reach shows a significant warming trend. The warming trend of Tuoketuo station is most obvious, the difference is 3.4 centigrade/10 years, the difference of Baotou is 2.7 centigrade/10 years, and the difference of Dengkou is minimum one, and which is 2.3 centigrade/10 years.

- In the last 50 years, the difference of the winter minimum and maximum air temperature is obvious, especially for the Ning-Meng reach, the winter accumulated minimum air temperature can reach -50 centigrade, and maximum is only about -3 centigrade, the difference of maximum and minimum is 3 to 7 times.
- The winter accumulated air temperature of the upstream stations is higher than that of the downstream stations, such as Tuoketuo is lower than Yinchuan as 15-16 centigrade and lower than Dengkou as 6-7 centigrade.

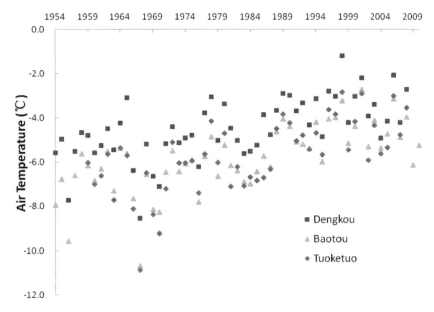

Figure 3.10 Mean winter air temperature variations at meteorological stations

Table 3.8 Statistical characteristic values of winter accumulated air temperature (Unit: ℃)

Item	Yinchuan	Dengkou	Baotou	Tuoketuo
Annual mean	-12.1	-21.2	-28.1	-27.7
Lowest	-32.0	-42.8	-53.5	-54.4
Highest	-2.9	-6.0	-13.3	-14.1
Difference (Centigrade/10 years)	2.4	2.3	2.7	3.4

More extreme weather events happened based on observation. From January 2002 to March 2010, the extreme events of the top three average temperature of ten-day of the same period in history have happened 24 times (Table 3.9), out of which the top ones have happened 8 times in Baotou station.

Every winter there was at least a ten-day period in which extreme events happened. Among the 24 times an extreme event occurred, air temperature on the high side happened 21 times, and on the low side it only happened 3 times. This also reflects the climate warming trend.

Table 3.9 Statistics of 10-day average air temperature since 2002

Year	Month	Ten-day	Average air temperature (centigrade)	History rank	Year	Month	Ten-day	Average air temperature (centigrade)	History rank
2002	1	First ten-day	-5.3	2nd high	2006	11	Last ten-day	-1.9	2nd high
		Middle ten-day	-3.8	1st high		1	Last ten-day	-7.5	3rd high
	2	Middle ten-day	0	2nd high	2007	2	First ten-day	-3	2nd high
		Last ten-day	1.2	3rd high			Last ten-day	1.3	2nd high
	12	Middle ten-day	-4.5	2nd high		3	Last ten-day	8.4	1st high
		Last ten-day	-17.4	3rd low		12	Last ten-day	-5.9	3rd high
2004	2	Middle ten-day	0.5	1st high	2008	1	Last ten-day	-16.4	3rd low
	12	First ten-day	-2.1	1st high		3	First ten-day	1.4	3rd high
		Middle ten-day	-4.8	3rd high			Middle ten-day	5.5	1st high
2005	3	Last ten-day	6.9	2nd high	2009	2	First ten-day	-0.4	1st high
	11	First ten-day	5.1	3rd high		11	Middle ten-day	-8.4	1st low
		Last ten-day	-1.9	2nd high	2010	2	Last ten-day	3	1st high

The daily average air temperature generally turns to negative stable at the beginning of November, and turns to positive stable by mid-March. Take the Baotou meteorological station as an example, the mean air temperature for the winter, usually changes between -10.7 centigrade and -2.7 centigrade, and the long term winter mean was -5.6 centigrade. On January 22, 1971, Baotou station had a -24.6 centigrade daily average air temperature, which was a record low.

Usually at the end of the first ten-day or the beginning of the middle ten-day of November which the average daily temperature starts turn to negative stably at all three meteorological stations in the reach of Inner Mongolia. The date when the average daily temperature of Dengkou station turns to negative stable is always a few days later than that of Baotou and Tuoketuo station. The date when the average daily air temperature turns to positive stable is almost the same at the 3 stations, generally the beginning of the middle ten-day of March. According to the statistics, the first freeze-up in the reach of Inner Mongolia will not appear until at least ten days later than that when average daily air temperature of Baotou station turns to stable negative, except for extreme special situations.

Air temperature inter-decadal variation characteristic and its influence on ice regime

According to the analysis of the inter-decadal variation average air temperature and ice regime characters of Ning-Meng reach, the air temperature inter-decadal variation characteristic and its influence on ice regime is the following:

- The winter average air temperature decadal variation tendency of the Ning-Meng reach is that in 1950s and 1960s the variation is low and since 1990 which increase significant, and in 1970s and 1980s which is close to the annual mean. The rate of change of 1950s and 1960s is negative anomaly, and the downstream is larger than that of upstream. Since 1990, the rate of change is positive anomaly and the downstream is close to that of upstream, and since 1990 the air temperature difference of that is small, therefore the effect on the ice regime is serious in 1950s and 1960s with long freeze-up days and thick layer of ice, and since 1990 with light ice regime.

- November and December is ice run and freeze-up period for the Ning-Meng reach, so the winter air temperature and its distribution play an important role for the ice regime. The winter air temperature variation total tendency of the last 50 years is low in 1950s and 1960s and 1970s for the upstream of the Ning-Meng reach, and since 1990 there is high, and for the downstream of the Ning-Meng reach, in 1960's is low, since 1990 is high, and the other decade is normal. Therefore for the ice regime, since 1990, the ice run and freeze-up time is later and the other decade is earlier and normal.

- The air temperature of February and March play an important role for time and situation of the river breakup period. The variation characteristic of the last 50 years is for the upstream of the Ning-Meng reach, in 1950s and 1990s is high, in 1970s and 1980s is low and 1960s is normal, for the downstream of the Ning-Meng reach, since 1990 is high, in 1950s and 1970s and 1980s is low and 1960s is also normal, Due to the breakup begins from the upstream, and finish at the downstream. Therefore the late breakup time mainly influenced by the air temperature of the downstream, since 1990 the breakup time is earlier than that of the multi-year average and at the other decade which is later than that of the multi-year average.

Air temperature variation and its influence on the ice regime

Using the winter air temperature data of the last 50 years, the correlation coefficient of average air temperature of November with the ice run time and freeze-up time and duration and freeze-up length is computed. The average air temperature of March for the Ning-Meng reach with breakup time respectively analysis and the air temperature variation and its influence on the ice regime leads to the following conclusions:

- The influence of air temperature variation on the ice run and freeze-up time: The calculated results show average air temperature of November for the Ning-Meng reach has close correlation with ice run and freeze-up time. The most correlation coefficient of the Ning-Meng reach is larger than 0.50, especially with the ice run time, most of them over 0.60, and confidence level over $a=0.001$, which indicates

66

the variation tendency of air temperature and ice run time is same. Which means the high air temperature, the late ice run and freeze-up time.

- The influence of air temperature variation on the freeze-up length and duration: Through calculation the correlation coefficient of monthly and winter average air temperature with freeze-up length and duration, which exists the significant negative correlation, especially for the Tuoketuo, the correlation coefficient of average air temperature and freeze-up length reaches from -0.69 to -0.71.

- The influence of air temperature variation on breakup time: The calculated results show average air temperature of March for the Ning-Meng reach with breakup time exist the significant correlation, which the Dengkou, Baotou, and Tuoketuo the correlation coefficient is between 0.64 and 0.78.

Calculation and analysis of air temperature variation on ice regime

Through the above analysis, we can preliminary realize the effect of the winter air temperature on the variation of ice regime. However, what is the relationship of air temperature variation and ice regime and how much effect on the ice regime should also be understood during the ice flood control. Furthermore, based on the statistical relationship between air temperature characteristics and ice regime, a relation formula could be inferred to analyse of the influence on ice regime of air temperature variation and other factors can be performed, as shown below.

The calculation formula

According to the variation character of the Ning-Meng reach and put into operation of the large water conservancy projects on the main stream, and let the Liujiaxia reservoir put into operation before 1968 as the natural environment period with less human activities influence, using the daily average air temperature characteristics data of Dengkou, Baotou and Tuoketuo from 1954 to 1967, considering the freeze-up time, average ice depth and freeze-up duration, selecting the air temperature factor which the correlation confidence at $a=0.05$ ($r>0.497$)to carry on the stepwise regression analysis, to set up the empirical relation, and then let the air temperature after 1968 to as the input into the relation formula, to calculate the ice regime characteristics of the natural conditions, and through the comparison of the former ice regime characteristics with the present real time observed ice regime data, to analyse the effect of air temperature variation and other factors variation on ice regime.

Through calculation the following relations were deduced:

$$y_1 = 231.096 - 3.712x_1 + 0.621x_2 \tag{3.1}$$

$$y_2 = 78.012 - 0.023x_5 - 0.3195x_6 \tag{3.2}$$

$$y_3 = 0.245 - 0.023x_3 + 0.00834x_4 \tag{3.3}$$

where, y_1, y_2, y_3 is freeze-up time, freeze-up duration and breakup time respectively, x_1 and x_2 is number of days which daily air temperature less than -5 centigrade and average intensity less than -10 centigrade of Dengkou respectively, x_3 and x_4 is average intensity above freezing point and number of days above -5 centigrade of Dengkou respectively, x_5 and x_6 is accumulated negative air temperature less than -5 centigrade and average intensity less than -7 centigrade of Dengkou resp.

Results and analysis

Using the air temperature characteristic values since 1968 as the input to the above relation formula, the ice regime characteristic values can be calculated, as exemplified in Table 3.10.

Based on the calculated results (Table 3.10), the discharge and air temperature data in the same period compared with the former ice regime, leads to the conclusions:

- Freeze-up time: Since 1970, the air temperature increase in November and December, especially in 1980s and 1990s, the average air temperature increase 1 to 2 centigrade, and postpone 2 to 5 days of the freeze-up time. At the same time, due to influence of the other factors, which forward 2 to 3 days of the freeze-up time, and through discharge contract analysis before and after the Liujiaxia Reservoir put into operation in 1986, can consider this mainly caused by the discharge decrease before the river freeze-up.
- Breakup time: Since 1970, due to the winter air temperature increase compared with 1950s and 1960s, especially for the Ning-Meng reach; since 1990s the average air temperature increased 1 to 3 degrees, which results in early river breakup time.

- Freeze-up duration: Since 1970, due to the air temperature during the freeze-up and breakup period is higher than that of the former, so the freeze-up time decreases, and together with the influence of the reservoir, the thermal breakup time is postponed and the freeze-up time increases compared with 1950s/1960s.

Table 3.10 Ice regime characteristic values and results of the Ning-Meng reach

Item	1954-1967			1968-1979			1980-1989			1990-2005		
	Freeze-up time (m.d)	Breakup time (m.d)	Freeze-up days (d)	Freeze-up time (m.d)	Breakup time (m.d)	Freeze-up days (d)	Freeze-up time (m.d)	Breakup time (m.d)	Freeze-up days (d)	Freeze-up time (m.d)	Breakup time (m.d)	Freeze-up days (d)
Observed	12.01	3.07	114.6	11.28	3.07	119.7	12.03	3.03	110.3	12.05	2.25	98
Calculated	12.01	3.07	114.6	12.01	3.05	111.5	12.05	3.01	109.0	12.06	2.23	102
Total variation				-3	+2	+8.1	0	-1.8	-4.3	-2	-2.6	-4.0
Air temp. influence				0	+3	-0.1	+2	-3.2	-5.6	-3	-1.9	-8.2
Other influence				-3	-1	+8.2	-2	+1.4	+1.3	+1	-0.7	+4.2

Inflow and outflow discharge variation

The tributaries in the Ning-Meng reach are all belong to seasonal rivers with small catchment area, for which the discharge is very low during the rainstorm flood period, and almost empty in the ice flood season. Therefore the discharge of the Ning-Meng reach is mainly influenced by the upstream inflow together with the intake water for irrigation. Before Longyangxia and Liujiaxia reservoir were put into operation, the inflow of the upstream basically adjusted naturally, and after that, the inflow of the upstream is adjusted by the reservoirs regulations.

Table 3.11 shows winter mean monthly discharge and total runoff at hydrometric stations of the upper Yellow River, the mean total runoff from November to March of 2000/2010 for Lanzhou is 7.52 billion m^3, and comparing with that of 1970-2000, which is 4% lesser, and for the hydrological stations in the Ning-Meng reach, the mean total runoff is about 10% lesser. And for the mean monthly discharge variation of Lanzhou station, comparison before and after 2000, besides November the discharge increase, the other months are all decrease, which the maximum decrease is 13% lesser than that in February, and then 9% lesser in January. Therefore, during the freeze-up period of the Ning-Meng reach the discharge of Lanzhou station decrease obviously in 2000-2010 comparing with that before 2000.

Table 3.11 Winter monthly discharge and runoff on the upstream of the Yellow River

Station	Period	Discharge (m³/s)					Runoff (billion m³)
		Nov.	Dec.	Jan.	Feb.	March	Nov.-March
Lanzhou	Mean of 1970-2000	839	596	553	518	505	7.86
	Mean of 2000-2010	870	563	505	448	493	7.52
	Anomaly（%）	4	-5	-9	-13	-2	-4
Shizuishan	Mean of 1970-2000	677	629	533	554	540	7.65
	Mean of 2000-2010	613	591	488	502	474	6.96
	Anomaly（%）	-10	-6	-9	-9	-12	-9
Bayangaole	Mean of 1970-2000	641	565	504	547	583	7.41
	Mean of 2000-2010	582	540	433	516	516	6.74
	Anomaly（%）	-9	-4	-14	-6	-11	-9
Sanhuhekou	Mean of 1970-2000	614	498	508	563	703	7.53
	Mean of 2000-2010	591	381	427	510	662	6.70
	Anomaly（%）	-4	-23	-16	-9	-6	-11
Toudaoguai	Mean of 1970-2000	541	426	463	528	762	7.10
	Mean of 2000-2010	476	319	316	435	814	6.17
	Anomaly（%）	-12	-25	-32	-18	7	-13

For the hydrometric stations of the Ning-Meng reach, the discharge from November to February, the decrease range of the lower stations is higher than that of the upper stations. The monthly discharge anomaly is -23%, -16% and -9% from December to February for Sanhuhekou station respectively, and for Toudaoguai station which is -25%, -32% and -18% respectively, especially for January, the monthly discharge negative anomaly of Toudaoguai is twice than that of Sanhuhekou, which illustrate the channel water storage increase from 2000 to 2010 comparing with that before 2000.

HUMAN ACTIVITIES

Reservoir regulation

Reservoir operation has a direct effect on the river discharge and water temperature during the ice season. At the same time, reservoir operation can influence the ice regime by changing sediment deposition and erosion of the channel. Reservoir operation changes depending at the moment of the year for the river ice run, freeze-up and breakup, after the Liujiaxia reservoir was put into operation in 1968. The annual mean freeze-up date was postponed comparing with the moment when freezing was happening before the construction and the operation of the Liujiaxia reservoir. After Longyangxia reservoir was put into operation in 1986, the discharge increased and taking into account the climate changes the date for freeze-up was postponed even further, such as Shizuishan, Bayangaole and Sanhuahekou is later with 20 days, 16 days and 8 days respectively comparing with that before 1968, and the Toudaoguai is earlier with 9 days comparing with that before 1986.

After the reservoirs were put into operation, the number of ice jams increased during the freeze-up period, and decreased during the breakup period for the Ning-Meng reach. The number of ice jams during the freeze-up period was 2, from 1950 to 1968, and 13 from 1969 to 2010. However, the annual mean number of ice jams during the breakup period is 13.6 from 1950 to 1968, and only 3.7 from 1969 to 1990. The ratio of thermal breakup to mechanical breakup is 1:1 from 1950 to 1968, and is 7:3 from 1969 to 1990. Although, reservoir regulation diminished the ice flood disasters, it produced new problems such as river sediment load increase rapidly and river flow capacity decrease almost one third of the previous one (the bank full discharge from 5000 m^3/s to 1500m^3/s).

Generally the dams usually lead to increase of sediment upstream of the dam and a decrease downstream. However, for the Ning-Meng reach of the Yellow River, the water volume mainly came from the river source area before Liujiaxia reservoir, the sediment came from the Loess plateau located at the downstream of the Liujiaxia reservoir, which the

71

sources are different. Then the reservoir regulation changes the natural flow to artificial one, most of the time is stable, another factor is the river of the Ning-Meng reach is flat, these are the factors to increase sediment load at the Ning-Meng reach.

Artificial diversion of the ice water from channel may effectively decrease the channel water storage increment of downstream, so that the dike pressure can be relieved. In the last 3 years, Sanshenggong dam is used for ice water diversion (Liu et al., 2011). Artificial ice water diversion helped reduce channel water storage increment around Sanhuhekou station, lower the risk of mechanical breakup and prevent ice flood disasters.

Structures along the river and agriculture irrigation

With the social and economic development along the river in the Ning-Meng reach, the bridges across the river increased. There are 3 railway bridges and 10 highway bridges between Bayangaole and Wanjiazhai. There are 12 pontoon bridges built in the last ten years. The structures across the river, including bridges, pontoons, construction piles and cofferdams can affect the ice formation and transport. There were 4 years continuously in the last ten years with first freeze-up position located near the railway bridge of Baotou from 2005 to 2008.

The first twenty days of November every year are still in the irrigation period in Ning-Meng irrigation area. The diversion and returning water in this area can effect on the process of freeze-up in the winters with earlier ice cover. The irrigation can lead to smaller discharge in channel and make the freeze-up date earlier. The returning water to channel after irrigation can lead to the discharge increase in the ice covered reach, so the low flow process appears twice at Taodaoguai and the channel storage usually gets larger than normal. Moreover, with the agriculture economy developing, there were massive reclaiming beach lands in flood plain area in Inner Mongolia which affected not only in flood season but also in ice season by intercept ice and water in the beach land, increase the channel water storage that released slowly during river breakup and impacted the process of ice flood peak, and decrease the discharge into downstream.

3.5 Summary

Tracking ice formation from observations and combining them with numerical model predictions for advanced warning requires proper understanding of all processes that play a role. However, it is not possible to make specific predictions because our physical understanding remains incomplete, thus the main challenge is how to accelerate the pace of discovery and bridge the major knowledge gaps. In the case of the Yellow River, ice floods impose a threat every year, which is why the YRCC is putting considerable effort in verifying theoretical formulations with actual field measurements in order to better understand the scientific mechanisms that play a role.

The formation, development and dissolution of the ice are mainly determined by the channel morphology, the hydrometeorological regime and human activity. The ice regime and human activities have impacted each other. The reservoir operation can influence the ice regime downstream by changing the discharge and water temperature. Some projects such as bridges and pontoons in channel can cause resistance to ice and water flows, and also influence the ice regime. The hydrological and meteorological data from 1950 to 2010 has been used for analyze the characteristics of ice regimes, especially after the Liujiaxia reservoir put into operation in 1986.

The air temperature of the Ning-Meng reach is below freezing point during the winter which last 5 months, and lately shows a significant warming trend. The warming trend of Tuoketuo station is most obviously, the difference is of 3.4 centigrade/10 years. Air temperature influences the stable freeze-up duration has short tendency, water temperature has increase tendency, and ice cover thickness has decrease tendency. The moment and date of ice run, freeze-up and breakup for river have been analysed, which can give the essential ice regime information for in which region and period is easy for initial ice run, freeze-up and breakup. Generally, for the Ning-Meng reach, the initial ice run date is around the middle and late November, and the average ice run date from 1970 to 1990 is on Nov.19 and that from 1990 to 2010 is on Nov.21, which is 2 days later. The river average freeze-up date is around the early December, from 1990 to 2010 is on Dec.3 and 1 days later compare with that from 1970 to 1990. The river breakup date is around the middle and late March, from 1990 to 2010 is on March 24 and 2 days earlier than that from 1970 to 1990.

The river average stable freeze-up duration is 117 days in the Ning-Meng reach. Comparing that from 2000 to 2010 with that of 1950-2010, the stable freeze-up duration is shorter by 3 days and has a short tendency. From 1990 to 1999, the mean ice cover thickness is only 55-58 cm, and in recent ten years from 2000 to 2010, which is 53-63 cm, and comparing with that before 1990, the mean ice cover thickness is thinner over 10 cm. The water temperature increases by about 0.7 centigrade from Shuizuishan to Bayangaole during November after 1986 as compared with the period when only Liujiaxia reservoir was operated. At Sanhuhekou station it increases by 0.3 centigrade, and at Toudaoguai it decreases by 0.5 centigrade. In March, the water temperature increases by 2.4 centigrade, 1.9 centigrade, 0.7 centigrade and 0.4 centigrade for Shuizuishan, Bayangaole, Sanhuhekou and Toudaoguai respectively. The main reason is that after Liujiaxia and Longyangxia reservoirs were put into operation, the discharge operated through the reservoir which has higher water temperature significantly affect that at Bayangaole during the river ice run, freeze-up and breakup period. In the same time on the reach from Sanhuhekou to Toudaoguai there is no evident effect of the reservoir operation, which is to be expected, because of the far away location of this reach from the reservoirs.

The mean channel water storage is about 1.24 billion m^3 from 1970 to 2010, and the maximum one is 1.91 billion m^3 which occurred in the ice flood period of 2004/2005, and the minimum one is only 0.45 billion m^3 which occurred in the winter of 1996/1997. And since 2000, the mean channel water storage has increased to 1.59 billion m^3. For the total tendency since 1990, besides the winter of 1996/1997, the channel water storage obviously increasing which comparing with that before 1990. Therefore, the total tendency of the channel water storage increased obviously since 1990.

The water level variation of Sanhuhekou and Toudaoguai stations have a tendency to increase during the freeze-up period, especially for the Sanhuhekou, where the water level increased 2 to 3 m since 1986. And the water level has the same increasing tendency for the breakup period, which for the Sanhuhekou, the historical first high water level occurred on March 20, 2008 was 1021.22 m, which was 0.41 m higher than the highest-ever level on record of the station, and leaded to dike-break. Due to the serious channel sedimentation and shrivel, the riverbed elevation increased and the main channel discharge capacity decreased, which illustrates the water level has the tendency to increase.

The ice flood mainly occurred at the reach from Bayangaole to Toudaoguai, the historical maximum ice flood peak is 3,500 m^3/s in 1968 and the minimum one is 1,000 m^3/s in 1958 and the mean ice flood peak is 2,192 m^3/s from 1950 to 2010. After Liujiaxia and Longyangxia reservoir put into operation, the ice flood peak and 10-day water amount of Toudaoguai increase gradually comparing with the natural conditions, and in the last years after Wanjiazhai reservoir put into operation in 1999, the ice flood peak increase but the 10-day water amount decreases.

The reservoir operation is an important influence on the evolution of ice regime. Reservoirs water regulation play an important role for changing the ice regime characteristics, which can influence ice regime by changing sediment deposition and erosion of the channel and directly influence ice regime characteristics, furthermore, proper reservoirs regulation can diminish the ice flood disasters. The channel water storage, flood peak and volume and water level variations show after 1986, there is tendency of shorter stable freeze-up duration, ascending for water temperature, and thinner of ice cover thickness, larger channel water storage, less flood peak oppose to larger flood volume, higher water level during the river breakup period.

With the development of society and economy, the ice regime characteristics have changed due to climate changes and human activities. At the same time, the ice characteristics changes have impact to local industry, agriculture and human life. With the analysis and understanding of the ice regime characteristics and its variation in the Ning-Meng reach, it is better to take measures to control and diminish the ice floods hazards, and useful for the building and operation of the ice flood modelling.

Chapter 4   Numerical Ice Flood Modelling

4.1   Model structure

In order to satisfy the requirements of safeguarding against ice floods and sufficiently utilize the limited water resources, a numerical ice flood model was developed, which is also one of the essential parts at YRCC. The model can be and currently is used to supplement the inadequacies in the field and lab studies carried out to help understand the physical processes of river ice on the Yellow River. The numerical ice flood model of the Ning-Meng reach consists of five modules (Figure 4.1): (i) pre-processing of meteorological data module which processes the observed meteorological data and numerical model output coming from CMA and the Hyper-computation Centre of YRCC; (ii) pre-processing of hydrological data from observations at hydrometric stations; (iii) initialization of ice regime module with observations from hydrometric stations; (iv) thermal balance calculation module, and (v) dynamic balance calculation module. These last two modules are the main part of the model and are coupled to each other during the model calculation.

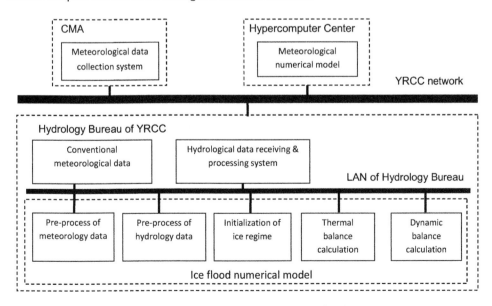

Figure 4.1 Components of YRCC numerical ice flood modelling

The modules logical structure of the numerical ice flood model of the Ning-Meng reach is presented in Figure 4.2. It consists of three main components: river hydrodynamics, thermodynamics, and ice dynamics modules. The model can be subdivided even further into the following modules: river hydraulics, heat exchange, water temperature, ice concentration distributions, ice cover formation, ice transport and ice cover progression, undercover deposition and erosion, thermal growth and decay of ice covers, and river freeze-up and breakup.

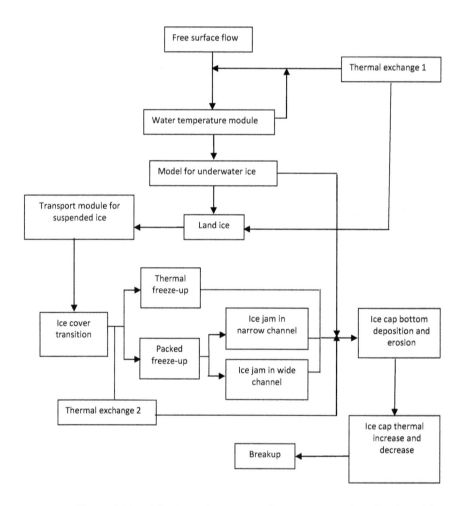

Figure 4.2 Modules logical structure of YRCC numerical ice flood model

The Flowchart of one-dimensional ice flood modelling is shown in Figure 4.3.

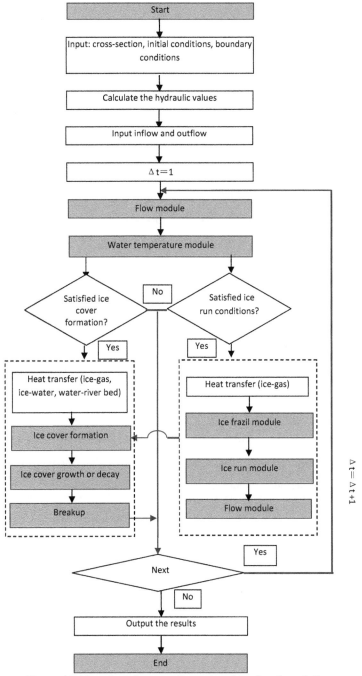

Figure 4.3 Flowchart for one-dimensional ice flood modelling

4.2 Air temperature forecast model

The numerical model results are combined with statistical methods to readjust the model outputs, so as to setup a medium and short-range air temperature forecast model for the Ning-Meng reach. The air temperature data include the surface air temperature data of the meteorological station and hydrological station of the Ning-Meng reach for the recent 40 years. First sets of 10-day forecast model data over the recent 5 years are selected, then statistical forecast methods are used to setup 1 to 10 days air temperature forecasts for Shizuishan, Linhe, Bayangaole, Sanhuhekou, Baotou, and Toudaoguai, then the results are used for numerical ice flood modelling of the Ning-Meng reach. This has proven to be a proper way to prolong the lead-time of ice flood forecasting.

4.2.1 Data sources

Meteorological models are an important tool for weather prediction and climate forecasting. Following the progress of numerical forecast technology and accuracy, at present the short-term (1 to 3 days) weather forecast has reached an acceptable standard. The future forecast using the rapid development of 10 to 15 days medium and short term numerical forecast model allows for outputs of the model that play an important role in weather prediction and climate forecast. For example, the Global Forecasting System (GFS) from the National Centers for Environmental Prediction (NECP), T639 from China Meteorological Administration (CMA), ECMWF from European Centre for Medium-Range Weather Forecasts, which can provide medium-range weather circulation forecast.

Considering the data conditions, here we select the T639 model forecast data as base to setup the statistical forecast model. The forecast includes height field, air temperature field, and wind field, with a time interval of 12 h, and time series is from 2009 to 2013. Surface observation data uses daily mean air temperature of 2 meteorological stations, Linhe and Baotou, and 4 hydrometric stations, Shizuishan, Bayangaole, Sanhuhekou, and Toudaoguai for a time period from 1974 to 2013.

4.2.2 Stepwise regression method

A stepwise regression method is used to setup the air temperature forecast model. The stepwise regression method consists in selecting k independent variables ($k \leq n$) from n independent variables, to fit the best linear multiple regression equation. The independent variables, which have no important influence on dependent variables should be eliminated out from the equation, otherwise should be keep inside the equation. The linear multiple regression equation is:

$$y = b_0 + b_1 x_1 + b_2 x_2 + b_3 x_3 + b_4 x_4 + b_5 x_5 + b_6 x_6 + b_7 x_7 + ... + b_i x_i \ (i=1,2,...,n) \tag{4.1}$$

where y is the dependent variable and , x_1, x_2, ...x_i are the independent variables, b_0, b_1, b_2, ...b_i are the regression coefficients of the equation. The steps are: (i) introduce one independent variable which influence y most from all variables, to setup regression equation only has one independent variable; (ii) based on (i) consider importing the second independent variable, to setup a regression equation which has two independent variables; (iii) consider whether or not to eliminate the independent variable in the regression equation; (iv) follow the above (ii) and (iii), import or eliminate one variable for the regression equation to only contain statistically significant independent variables until one cannot import and eliminate the independent variable in the regression equation.

4.2.3 Forecasting factor selection

In winter, the air temperature variation is mainly influenced by cold air mass moving, surface wind field, 500 hPa trough and ridge moving etc. So selecting 500 hPa geopotential height, 850 hPa air temperature, surface zonal wind and meridional wind, 2 m air temperature, and 30-year daily mean air temperature as forecast factors, and consider the air temperature variation continuous, so select previous daily mean air temperature as 24 h forecast factor. The selected forecasting factors and their meaning are shown in Table 4.1. Based on the above selected factors, the daily air temperature linear regression equations are setup for different meteorology and hydrology stations, separately.

Table 4.1 The selected forecast factors

Factor	Meaning
x_1	500 hPa geopotential height (10 dgpm)
x_2	850 hPa air temperature (centigrade)
x_3	Surface zonal wind (m/s)
x_4	Surface meridional wind (m/s)
x_5	2 m air temperature (centigrade)
x_6	30-year daily mean air temperature (centigrade)
x_7	Previous daily mean air temperature (centigrade)

4.2.4 Validation of air temperature forecasting model

By using the above 1 to 10 days forecast data from T639, which consist of observed daily mean air temperature data from 2009 to 2012, the daily air temperature linear regression equations for different meteorology and hydrology stations recorded in winter periods, a system of 60 equations was generated (Appendix A). The data from 2012 to 2013 and the 50 equations are used to carry on the application and validation. According to the CMA (China Meteorological Administration) method, the standard air temperature forecast accuracy for the previous three days is ±2 centigrade, and the latter 7 days is ±2.5 centigrade.

Figure 4.4 and Figure 4.5 show the 24 hrs daily average air temperature comparison of forecast based on observed data at Sanhuhekou and Toudaoguai from 2012 to 2013. Table 4.2 shows 1-3 days, 4-7 days and 8-10 days forecast accuracy, statistical averaged results from 2012 to 2013. The monthly result of the forecast are better is in January, and the in November. The result for the 1-3 days forecast in February is not so good; and the 4-7 days and 8-10 days forecast results in March are even worse. The analysis of the forecast lead-time show that the 1-3 days and 4-7 days forecast accuracy is better than that of 8-10 days; the 1-3 days and 4-7 days forecast accuracy of Sanhuhekou in January has reached 88.7% and 84.8% respectively; while the 1-3 days and 8-10 days forecast accuracy of every stations in January reached over 70%. From the forecast stations, the forecast results of Baotou and Sanhuhekou are better, the air temperature average forecast accuracy from

November to March is larger than 60%, and 1-3 days and 4-7 days is larger than 69%. The 1-3 days and 4-7 days forecast accuracy of other three stations are all over 65%. So most of the forecast average absolute error of generated daily air temperature linear regression equations for 5 stations separately are less than 2 centigrade, so air temperature has a forecast ability of 10 winter days.

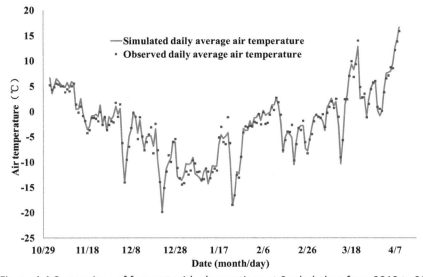

Figure 4.4 Comparison of forecast with observations at Sanhuhekou from 2012 to 2013

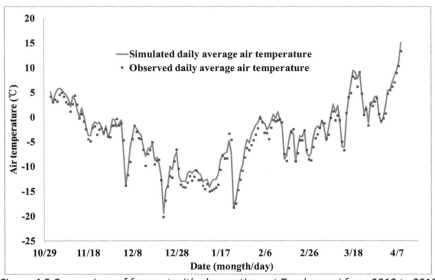

Figure 4.5 Comparison of forecast with observations at Toudaoguai from 2012 to 2013

Table 4.2 1-3 days, 4-7 days and 8-10 days forecast accuracy results from 2012 to 2013

Forecast period (d)	Month	Linhe (%)	Baotou (%)	Bayangaole （%）	Toudaoguai (%)	Sanhuhekou (%)
1-3	Nov.	68.1	77.1	68.2	73.0	80.0
	Dec.	65.3	70.9	61.4	72.6	68.6
	Jan.	75.3	81.6	69.7	71.3	88.7
	Feb.	68.9	65.3	63.5	61.7	61.5
	March	65.7	76.5	60.5	66.0	68.7
4-7	Nov.	63.8	69.2	69.2	67.9	65.8
	Dec.	59.2	72.9	60.5	66.3	70.4
	Jan.	78.2	82.7	74.2	73.4	84.8
	Feb.	65.6	71.9	62.9	70.6	63.9
	March	59.9	67.4	57.4	58.2	61.1
8-10	Nov.	59.5	75.5	65.6	69.5	72.8
	Dec.	46.1	58.9	48.3	55.6	58.3
	Jan.	59.7	74.2	51.1	58.6	65.1
	Feb.	55.4	53.6	54.8	56.5	55.3
	March	43.3	52.0	44.3	44.8	51.9

4.3 Numerical ice flood modelling

4.3.1 Mathematical representation

Continuity and momentum equations

The governing principles for flood simulation in rivers are conservation of volume, momentum and energy. These physical conservation principles were developed by Isaac Newton (1687), who introduced a clear mathematical formulation. By applying these principles to rivers, De Saint Venant (1871) formulated the mathematical equations for modern river flow simulations. The one-dimensional equations for continuity and momentum (as defined by De Saint Venant) can be written as:

$$\begin{cases} \dfrac{\partial A}{\partial t} + \dfrac{\partial Q}{\partial x} = 0 \\[2mm] \dfrac{\partial Q}{\partial t} + \dfrac{\partial (Q^2/A)}{\partial x} + gA\left(\dfrac{\partial Z}{\partial x} + S_f \right) = 0 \end{cases} \tag{4.2}$$

where, t is time (s); x is distance (m); Q is discharge (m³/s); A is net flow cross-sectional area (m²); g is gravity acceleration (m/s²); Z is water level (m); S_f is river bed slope. The solution of these equations is obtained numerically by applying the Preissmann scheme on the discretized form of these equations.

Ice heat exchange

The temperature of the winter water body depends on the heat exchange of water body and surrounding environment. The channel water body heat exchange including four parts (Figure 4.6): water and atmosphere heat exchange ϕ_{wa}; water and ice heat exchange ϕ_{wi}; ice and atmosphere heat exchange ϕ_{n}; water and river bed heat exchange ϕ_{wb}. Among them the heat exchange between water and atmosphere plays a leading role on the heat exchange of water body and surrounding environment, which contains the heat source coming from solar short wave radiation φ_{sn}, and atmosphere long wave radiation φ_{an}; the heat loss for heat transfer φ_c from incoming flow, water body long wave reflective radiation φ_c, and evaporation φ_e. The calculation methods and formula of the above mentioned ice heat exchange can be found in Shen and Chiang (1984).

Water temperature simulation

For fully-mixed rivers, the water temperature changes along the flow direction can be described by a one-dimensional diffusion equation (Lal and Shen, 1991):

$$\frac{\partial}{\partial t}(\rho C_p A T_w) + \frac{\partial}{\partial x}(Q \rho C_p T_w) = \frac{\partial}{\partial x}(A E_x \rho C_p \frac{\partial T_w}{\partial x}) - B_0 \phi_T \tag{4.3}$$

where, B_0 is the river width between border ice (m); T_w is the water temperature (centigrade); C_p is the specific heat of water (4,148 J/(kg centigrade)); ρ is water density (1,000 kg/m^3); E_x is the dispersion coefficient; ϕ_T is the net rate of heat loss per unit surface area of the river.

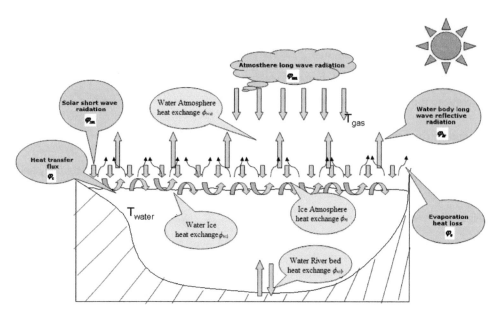

Figure 4.6 Heat exchange chart

Ice run concentration simulation

When water temperature drops below the freezing point, frazil ice will be generated. Using the one-dimensional diffusion equation to rewritten the cross-section averaged ice concentration distribution as following (Shen and Chiang, 1984):

$$\frac{\partial}{\partial t}(\rho_i L_i C_i A) + \frac{\partial}{\partial x}(Q \rho_i L_i C_i) = \frac{\partial}{\partial x}(A E_x \rho_i L_i \frac{\partial C_i}{\partial x}) + B_0 \phi_T \tag{4.4}$$

where, ρ_i is the ice density (917 kg/m^3); C_i is the ice run concentration; L_i is the latent heat of fusion (4,200 J/(kg centigrade).

Ice transport and other ice related simulation

Ice transport mainly calculates the physical process from underwater ice to land ice and thin ice before ice cover is formed. The equation of ice frazil motion can be applied to both open river and ice covered river, while the surface ice discharge double-layer mode can be used in open rivers, its function mainly divides all ice discharge into water surface ice and floating ice transport according to the determined distribution coefficient in the double layers analysis. Here, the convention-diffusion equation between ice run and floating ice used to simulate the ice transport procedure, which can reflect the actual conditions.

And for the other ice related modules of the ice flood modeling, the skim ice formation is based on the empirical equation, developed by Matousek (1984); growth of the border ice is based on Michel et al. (1982); frazil ice along the river channel is described by the mathematical model defined by Shen and Chiang (1984) and the ice dynamics is described by the static border of the ice formation, which is defined by Svensson et al. (1989) in form of a critical value.

4.3.2 Changes in mathematical representation

Adjusted continuity and momentum equation

The river dynamics representation using the De Saint Venant Equations, is generally used for open water flows, however, for the ice covered river, due to the ice in the water and ice covered above the water, the flow becomes two phase flow and should consider the thermodynamics. Furthermore, especially for the Ning-Meng reach of the Yellow River basin, for which the river channel is flat and wide, during the winter freeze-up period, with the water level increase, a large amount of water goes into the floodplain and become the water storage and ice cover there. The annual average water storage is about 1.24 billion m^3 from 1970 to 2010, which is a solid reservoir, and during the breakup period of the river, there are releases from the floodplain and the melting ice flows back into the channel to form the ice flood. Considering the winter situations of the Ning-Meng reach, for the continuity equation, added channel water storage terms which including the ice cover term

$(\frac{\partial Nt_i'}{\partial t})$ and floodplain water and ice term (q_l) to keep the mass balanced. And for the momentum equation, added ice cover friction term (gAS_{ix}) to keep the momentum balance.

$$\begin{cases} \dfrac{\partial A}{\partial t} + \dfrac{\partial Q}{\partial x} = \dfrac{\partial Nt_i'}{\partial t} + q_l \\ \dfrac{\partial Q}{\partial t} + \dfrac{\partial (Q^2/A)}{\partial x} + gA\left(\dfrac{\partial Z}{\partial x} + S_f + S_{ix}\right) = 0 \end{cases} \tag{4.5}$$

where, N is ice concentration (m); t_i' is submerged ice depth (m), ordinary t_i' is 0.9 of the ice thickness; q_l is unit floodplain water and ice (m^2/s); S_f is river bed slope, $S_f = n_c^2 \dfrac{|u|u}{R^{4/3}}$; n_c is roughness of river bed and ice cover, namely composite Manning's coefficient (Larsen, 1969), $n_c = \left[(n_i^{3/2} + n_b^{3/2})/2\right]^{2/3}$; n_i is ice cover roughness; n_b is river bed roughness; u is cross-section average velocity (m/s); R is hydraulic radius (m); S_{ix} is ice cover slope.

Channel water storage terms

Channel water storage calculation is mainly based on the observed discharge of hydrometric cross-sections. According to the water balance principle, channel water storage terms can be calculated as:

$$\frac{\partial Nt_i'}{\partial t} + q_l = \sum_{\Delta t} (Q_i - Q_j)/L \tag{4.6}$$

where, Q_i, Q_j are inflow and outflow discharge of the river reach (m^3/s); Δt is time period (s), commonly takes 1 day; L is the river reach length (m).

Simplified water temperature simulation

For the river reach when the water temperature varies little along the river, $\dfrac{\partial T_w}{\partial x} = 0$, then Eq.(4.3) can be simplified as follows:

$$\frac{\partial}{\partial t}(\rho C_p A T_w) = -B_0 \phi_T \tag{4.7}$$

The ϕ_T may be approximated by the rate of surface water atmosphere heat loss, ϕ_{wa}.

$$\phi_T = k_{wa}(T_w - T_a) \tag{4.8}$$

where T_a is air temperature (centigrade); k_{wa} is heat exchange coefficient between water and atmosphere (W/m²).

The simplified Eq.(4.7) and Eq.(4.8) may be used during the river ice run and freeze-up period, when the water temperature remains freezing point or varies little along the river, however, for the river breakup period, due to the water temperature increase and varies a lot along the river, it is better to use the former one to calculate the water temperature.

Simplified ice run concentration simulation

Eq.(6.4), due to the magnitude of the dispersion terms ($\dfrac{\partial Q}{\partial x}, \dfrac{\partial C_i}{\partial x}$), which are very low comparing with the other terms and can be neglected, is rewritten as:

$$\frac{d}{dt}(\rho_i L_i C_i A) = B_0 \phi_T \tag{4.9}$$

After the ice run starts, the water temperature equals to the freezing point, then Eq.(5.9) can be simplified as:

$$\Delta C_i = \frac{B_0 k_{wa}}{A \rho_i L_i} \Delta T_a \tag{4.10}$$

From Eq.(4.10), using the previous time step ice concentration and air temperature data (observed and forecasted one) to calculate the ice concentration variation.

River freeze-up

After river ice run, following the air temperature decreases, the water body loss heat gradually, together with the ice run concentration increase, and finally result in the river freeze-up. The ice frazil mainly affected the frictions and freezing force with other ice frazil, and drag force of water flow, and wind force. When the resistance forces increase larger than the drag forces, then the river starts to freeze-up. Though the river freeze-up mainly influenced by the ice frazil concentration and ice transport capacity at the cross-section. The ice frazil concentration mainly has the relation with the air temperature, and the ice transport capacity at the cross-section mainly has the relation with the discharge and river course conditions. Take the C_{max} as the river course maximum ice transport capacity, if the river course ice transport capacity C_i is greater than or equal to C_{max}, then occurs the river freeze-up. Then the river course ice transport capacity mainly affected by discharge, river course conditions and negative air temperature, which is:

$$C_i = f(Q, T_{a-}, RC) \tag{4.11}$$

where T_{a-} is accumulative daily negative air temperature (centigrade), measured from the day when air temperature turns negative; RC is river course conditions.

C_i is direct proportion to discharge, larger discharge means larger flow velocity, and larger ice transport capacity. C_i is inverse proportion to accumulative negative air temperature, which larger accumulative negative air temperature means more freezing force and high possibility of river freeze-up. Thus the river course maximum ice transport capacity can be described as follows:

$$C_{max} = \delta \frac{Q}{\sum T_{a-}} \tag{4.12}$$

where δ is experience parameter, which indicates the influence of river course conditions etc. on the ice transport capacity, and can use the observed data to calibrate. Then using the Eq.(4.12), to setup the river freeze-up judgment:

$$\delta \geq \frac{C_i \left| \sum T_{a-} \right|}{Q} \tag{4.13}$$

Forecasted air temperature data and observed discharge data are used to calculate the δ value, and to judge the river freeze-up or not. Normally, from experience if $C_i \geq 0.75$, then consider the river as freeze-up.

River breakup

At the end of winter or at the beginning of spring, when the air temperature rises to above freezing point, the ice cover begins to melt gradually and its strength decreases. Following the runoff increasing, the water temperature increase and the ice cover turn thinner and colour changed, comes across the better hydro-thermodynamic conditions, the river will be broke up. The melting of the ice cover begins from the land side, and gradually enable the ice cover out of bank, when water level increase visibly, with proper hydraulic force and wind pulling exerted on the ice cover which greater than resistance strength of that, the ice cover will moving or destroyed. The condition for ice cover broken can use the inequality as follows:

$$\xi h_i \leq f(H, \Delta H) \tag{4.14}$$

where ξ is ice cover intensity (Pa); h_i is ice cover thickness (m); ΔH is water level increment (m). And due to $H=f(Q)$, thus the above Eq.(6.14) can be rewritten as:

$$\xi h_i \leq f(Q, \Delta Q) \tag{4.15}$$

where ΔQ is discharge increment (m³/s).

The ice cover intensity and depth are the thermal factors that prevent the ice cover to broke, which is mainly related to air temperature. The Q and ΔQ represent drag force and lifting force which the flow on ice cover respectively, thus these two factors are the hydraulic factors to impel the ice cover broken. When the factors of impel ice cover broken are larger than that of prevent ice cover broken, then the river breakup. The river ice-breakup phenomena involve the following parameters: river discharge (hydraulic factor) Q; air temperature (thermal factor) T_{a+}; and ice thickness (ice cover intensity) h_{i+} which combined give an empirical criterion for river ice-breakup:

$$Q \geq \frac{9.5\alpha_* h_{i+}^{0.95}}{\sum T_{a+} + 1} \tag{4.16}$$

where h_{i+} is ice thickness when air temperature varies from negative value to positive one (m); α_* is river breakup coefficient (generally $\alpha_*=4$ for mechanical breakup and $\alpha_*=22$ for thermal breakup); $\sum T_{a+}$ is the accumulative daily positive air temperature, measured from the day when air temperature turns positive.

After Longyangxia reservoir was put into operation in 1986, with the influence of reservoir regulation and air temperature increase, the discharge of the river breakup key period decreases, thus the effects of hydraulic factors decrease, and reversely effects of thermal factors increase gradually. Based on this situation of the Ning-Meng reach, rewrite the Eq.(5.16) as

$$RB = \frac{Q^{0.8}(\sum T_{a+})^{1.1}}{D_r h_{i+}}$$
(4.17)

where, D_r is river channel conditions coefficient, which can use the observed data to calibrate. Using the results of air temperature forecast and observed discharge data, to calculate the RB value, and then to judge the river breakup or not.

4.3.3 Program design and real time calibration

The program designed of ice flood modelling consists of one main program and 14 subroutines detailed as boxes in Figure 4.6 and functionally defined in Table 4.4. Since a mathematical model is the abstract generalization of the real physical phenomena, errors inevitably exist on the model structure in the process of calculation, and also influenced by the numerical discretization error, rounding error and truncation error, all of which making the results of model calculations deviate from the 'true' value. Based on the principle of ensemble Kalman filter, the real time observation data of water level and discharge is used to improve the prediction accuracy of the hydrodynamic mathematical model, and realize the inversion of ice cover roughness parameters. To formulate the state spatial expression for dynamics variables, such as water level and discharge, to implement reasonable error disturbance for boundary conditions, such as water level and discharge, and using the ensemble Kalman filter algorithm to obtain the optimum outcomes. Meanwhile, taking the roughness parameters as state variables to simulate the dynamic process, and taking the updated model state as the initial condition of the next time step, to optimize the model prediction results.

Table 4.3 Subroutines function of the ice flood modelling

Name	Function
FLOW_PSM	To solve the hydrodynamics module
Roughness_Ice	To identify ice cover and river channel roughness
Cross_Heat_Exchange	To calculate heat exchange between water body and its environment at freeze-up or unfreeze-up conditions
Cross_Temperature	To calculate water temperature
Cross_Ice_Frazil	To simulate the growth and transport of ice frazil
Ice_Transport	To simulate ice movement, surface ice density variation and ice cover depth variation
Skim_Ice	To simulate the skim ice formation
Border_Ice	To simulate the border ice formation and amount
River_Freezeup	To judge each cross-section freeze-up or not
Ice_Cover_Progression	To judge the ice cover development type
Ice_Cover_Juxtaposition	To simulate ice cover progress as thermal freeze-up
Ice_Cover_narrowjam	To simulate ice cover progress as packed freeze-up for narrow-river
Ice_Cover_widejam	To simulate ice cover progress as packed freeze-up for wide-river
Ice_Growth	To simulate ice cover growth and erosion
River_Break	To judge each cross-section breakup or not

4.3.4 Framework of two-dimensional ice flood modelling

Based on mathematical representation presented in previous section, the continuity and momentum equation for the two-dimensional ice flood modelling should be changed to reflect the behaviors of the Ning-Meng reach. Here, for the continuity equation, a channel water storage term is added. This term includes the ice cover term ($\frac{\partial Nt_i'}{\partial t}$) and floodplain water and ice term (q_i) to keep the mass balanced. For the momentum equation, an ice cover friction term is added ($-ghS_{ix}$ and $-ghS_{iy}$ for x-direction and y-direction respectively) to keep the momentum balanced.

The mass continuity equation:

$$\frac{\partial h}{\partial t} + \frac{\partial (hu)}{\partial x} + \frac{\partial (hv)}{\partial y} = \frac{\partial Nt_i}{\partial t} + q_i \qquad (4.18)$$

x-direction momentum equation:

$$\frac{\partial (hu)}{\partial t} + \frac{\partial \left(hu^2 + gh^2/2\right)}{\partial x} + \frac{\partial (huv)}{\partial y} = v_t\left[\frac{\partial^2 (hu)}{\partial x^2} + \frac{\partial^2 (hu)}{\partial y^2}\right] - gh\left(S_{ox} + S_{fx} + S_{ix}\right) \qquad (4.19)$$

y-direction momentum equation:

$$\frac{\partial (hv)}{\partial t} + \frac{\partial (huv)}{\partial x} + \frac{\partial \left(hv^2 + gh^2/2\right)}{\partial y} = v_t\left[\frac{\partial^2 (hv)}{\partial x^2} + \frac{\partial^2 (hv)}{\partial y^2}\right] - gh\left(S_{oy} + S_{fy} + +S_{iy}\right) \qquad (4.20)$$

where h is vertical mean water depth (m); u, v is x and y direction velocity (m/s); v_t is turbulent viscosity coefficient; S_{ox}, S_{oy} is x and y direction river bed gradient respectively, its value is $\begin{pmatrix} S_{0x} \\ S_{0y} \end{pmatrix} = \begin{pmatrix} \partial Z_b/\partial x \\ \partial Z_b/\partial y \end{pmatrix}$; S_{fx}, S_{fy} is x and y direction friction slope, its value is

$$\begin{pmatrix} S_{fx} \\ S_{fy} \end{pmatrix} = \begin{pmatrix} n^2 h^{-4/3} u\sqrt{u^2 + v^2} \\ n^2 h^{-4/3} v\sqrt{u^2 + v^2} \end{pmatrix};$$

S_{ix}, S_{iy} being the ice cover friction along x and y direction respectively.

For the given governing equations, implicitly it is assumed that the turbulent viscosity coefficient is the same along x and y direction. The model uses unstructured triangle units and hybrid unit of triangle and square, created by a grid generator. Finite Volume Method (FVM) is mainly used for solving the discretized equations. In order to save the calculation time, the Multi Point Interface (MPI) parallel computing algorithm is used, as well as data partitioning methods, to realize the parallel analysis for two-dimensional river ice flood model.

The Flowchart of two-dimensional ice flood modelling is shown in Figure 4.7.

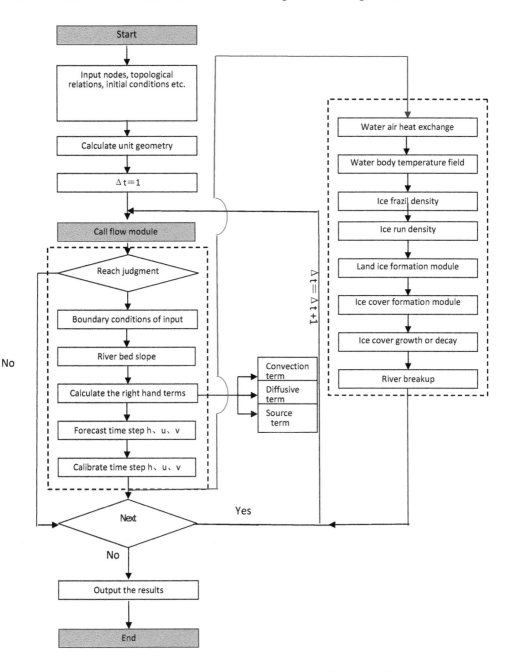

Figure 4.7 Flowchart for two-dimensional ice flood modelling

4.4 Parameters determination

The model could be determined under different ice regime conditions, it means that some parameters could be calibrated under open water condition, such as Manning coefficient of river bed and heat exchange coefficient between water and atmosphere; some parameters could be calibrated under ice condition, such as Manning coefficient under ice cover, ice roughness, and decay constant. Hence the model parameters determination procedure is divided into 'under open water condition' and 'under ice condition' to determine the different parameters.

4.4.1 Ice cover roughness

The ice cover roughness varies at different times, generally considering the freeze-up period, the roughness reaching its highest value, and at the end of the winter, reaching the lowest values (Nezhikhovskiy, 1964):

$$n_i = n_{i,e} + (n_{i,i} - n_{i,e})e^{-\alpha_n T} \tag{4.21}$$

where α_n is decay coefficient, the value of α_n is different according to cool winter, warm winter and mild winter as well as the frozen level, which the mean air temperature of the coldest 2 months less than -12 to -15 centigrade is cold winter, from -7 to -11 centigrade is mild winter and higher than -5 to -6 centigrade is warm winter. The value of parameter α_n which based on Nezhikhovskiy (1964) can be seen in Table 4.4. $n_{i,i}$ is the initial ice cover roughness, which is the one at the beginning of the river freeze-up, its value increase with the ice cover depth increase, its value which based on Nezhikhovskiy (1964) is shown in Table 4.5; $n_{i,e}$ is the ice cover roughness at the end of ice covered period or the frazil disappeared period, its value basically keeps as constant as 0.008 to 0.012; T is the days after the river freeze-up. When the river ice cover has formed, the friction slope is:

$$S_f = n_c^2 \frac{|u|u}{R^{4/3}} f \tag{4.22}$$

where: $f = \lambda(1+F)^{4/3} + \dfrac{(1+\lambda)(1+F)^{4/3}}{\left[\mu+(1-\mu)(1+F)^{3/2}\right]^2}$; $F = (n_i / n_b)^{3/2}$; μ is the percentage of smooth

flow area; λ is ratio of frozen river reach to total river reach (when $\lambda=1$ then the above Eq.(4.22) is the channel resistance calculation formula of the total frozen river reach).

Table 4.4 Value of parameter α_n (based on Nezhikhovskiy, 1964)

Type	Channel ice regime		
	More lead	Less lead	No lead
Cold winter	0.005	0.01	0.02
Mild winter	0.023	0.024	0.025
Warm winter	0.05	0.04	0.03

Table 4.5 Initial ice cover roughness value of $n_{i,i}$ (based on Nezhikhovskiy, 1964)

The initial depth of ice frazil accumulated formed ice cover (m)	Flow component		
	Loose frazil	Frozen frazil	Frozen ice
0.10	-	-	0.015
0.30	0.010	0.013	0.040
0.50	0.010	0.020	0.050
0.70	0.020	0.030	0.060
1.00	0.030	0.040	0.070
1.50	0.040	0.060	0.080
2.00	0.040	0.070	0.090
3.00	0.050	0.080	0.100

Here, ice cover roughness is computed using Manning equation and the observed hydrometric data for Sanhuhekou station. The results of these computations are shown in Figure 4.8, where ice cover roughness calculated with Eq.(4.21) is represented as variation in time along with the observed roughness in the field. The good match between the predicted roughness and the calculated one confirms that Eq.(4.21) can be used to calculate the ice cover roughness in the ice flood modelling of the Ning-Meng reach.

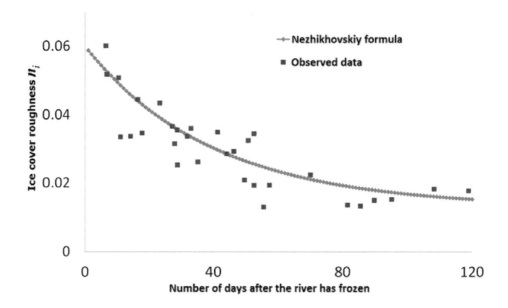

Figure 4.8 Ice cover roughness variation at Sanhuhekou

4.4.2 Ice heat exchange

In general in order to calculate the water and atmosphere heat exchange there is a need for more continuous observed meteorological data, which sometimes it is difficult to obtain. Ashton (1986) proposed a linearized approximation formula to simplify that, which uses the difference of water temperature and atmosphere temperature times heat exchange coefficient, and suggested for the northern United States this coefficient to take the value of 20 W/m^2 centigrade. In order to simplify the calculation processes, the coefficient for the Ning-Meng reach needs to be determined. Based on the winter half year meteorological conditions and data of Dengkou, Baotou, and Tuoketuo meteorological stations, for the open water reach without ice cover, the water atmosphere heat exchange calculation results are shown in Table 4.6, in which the heat balance theory and formula are used to calculate the river reach water body heat loss is 110.98 W/m^2, 111.57 W/m^2, and 112.78 W/m^2 respectively, which means the water body loss heat to atmosphere, to determine the water atmosphere heat exchange coefficient is 15.32 W/m^2 centigrade (with Standard Deviation s=23.2, Correlation Coefficient r=0.89) for the Ning-Meng reach.

Table 4.6 Water and atmosphere heat exchange calculation results (Unit: W/m^2)

Station	Water body net absorbed short wave radiation φ_{sn}	Atmosphere long wave radiation φ_{an}	Long wave reflective radiation of water body φ_{br}	Heat loss due to water surface evaporation φ_{e}	Heat transfer flux φ_{c}	Water atmosphere heat exchange ϕ_{wa}
Dengkou	72.32	252.27	-314.65	-72.54	-48.38	-110.98
Baotou	69.65	244.47	-306.84	-66.97	-51.88	-111.57
Tuoketuo	65.34	236.12	-296.23	-63.37	-54.64	-112.78

Table 4.7 Heat exchange calculation results during the freeze-up period (Unit: W/m^2)

Item	YC 68	YC 69	YC 70	YC 71	YC 72	YC 73	YC 74	YC 75	YC 76	YC 77
Water and river bed heat exchange ϕ_{wb}	9.7E-4	9.7E-4	9.7E-4	9.7E-4	9.7E-4	9.7E-4	9.7E-4	9.7E-4	9.7E-4	9.7E-4
Water and ice cover heat exchange ϕ_{wi}	-	-	-	-15.84	-20.09	-28.97	-25.02	-23.43	-34.77	-35.21

Note: "-" means unfreeze-up.

Table 4.7 shows the calculated heat exchange values between the water body and river bed and ice cover during the freeze-up period, which means after the ice cover formed, the turbulence heat transfer from the moving water body to ice cover, which as the heat generated by the friction of moving water body with ice cover and influence the ice cover depth. Here YC 68-YC 77 represents the locations of the cross-sections as defined in Figure 3.4. From the calculated values, after the ice cover formed, the temperature gradient between water body and ice cover bottom is not very large, so the heat exchange value generated by friction is very low comparing with the solar heat radiation. The heat exchange between riverbed and water body is very low, can be neglected comparing with the other heat exchange.

4.4.3 Water temperature

From above mentioned Section 4.3.2, the key of water temperature simulation is to calibrate heat exchange coefficient. And the heat exchange coefficient relationship with air temperature is

$$k_{wa} = ae^{bT_a}$$

(4.23)

Here, a and b are parameter which can be calibrated using the observed data. Table 4.8 shows the value of a and b at different hydrometric cross-sections.

Table 4.8 Calculated heat exchange coefficient parameters of water and atmosphere

Item	Shizuishan	Bayankaole	Sanhuhekou	Toudaoguai
a	4.7	5.0	8.0	8.0
b	0.02	0.05	0.05	0.05

4.4.4 River freeze-up and breakup criteria

The river freeze-up judgment mentioned in Section 4.3.2 is based on the actual conditions of the Ning-Meng reach, when $\delta \geq 0.012$ or the ice run concentration C_i is larger than 75%, the river will freeze-up. For the river breakup criterion mentioned in Section 4.3.2, the observed data are used to calibrate leading to Shizuishan, Bayangaole, Sanhuhekou, Baotou and Toudaoguai values of 1.0, 0.8, 1.2, 1.1, and 0.9 resp. When the value of RB is greater than 0.35, the river at this cross-section is broken up.

4.5 Test case

In order to test the results of ice flood model, due to the available cross-section observation data and other hydrological and ice regime data, the reach (Figure 3.4) from Zhaojunfen

(Yellow River Cross-section which short as YC 68) to Huajiangyingzi (YC 77) was selected. The river length is about 22.1 km as the test reach for the ice flood model testing, including the design of ice regime, meteorological conditions, simulate the generation and disappearance procedure of river ice. The boundary conditions for calculation are determined as follows:

o River reach: from Zhaojunfen (YC 68) to Huajiangyingzi (YC 77) which close to Baotou City;

o Topography data: the observed cross-section data after the flood period in 2012;

o Initial conditions: average freeze-up discharge 600 m³/s, water temperature 15.0 °C;

o Meteorological conditions: and the total solar radiation using the winter half year average radiation 2343.0 MJ/m², and sunshine percentage using the 40 years average value of 0.6273 in winter of the Yellow River basin, the air average humidity using 0.6, cloud coverage using the global average value of 0.6, the wind speed using 2 m/s at 2 m above the water surface, the air temperature at 2 m above the water surface as -10.0 °C;

o Input conditions: the discharge of input cross-section YC 68 as the average discharge 600 m³/s during the freeze-up period, the water temperature takes 0.0 centigrade, and ice run concentration as 0.0, the surface ice run concentration as 0.0, and the duration of these conditions lasting 30 days;

o Output conditions: cross-section Huajiangyingzi (YC77) using the rating curve (Fig. 4.9).

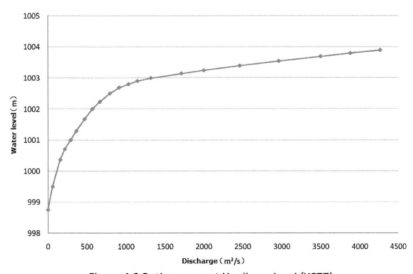

Figure 4.9 Rating curve at Huajiangyingzi (YC77)

Water temperature variation and frazil ice generation procedure

From Figure 4.10, with the influence of heat exchange and input water body temperature, the water temperature of YC 77 decrease from 15 centigrade and after 25 hours it reaches freezing point. Due to the water body loss heating takes place fast before the ice cover formed, the water temperature decrease to below freezing point gradually, and following the temperature gradient between water body and atmosphere decrease, the water temperature variation tends to stable, and keeps at -0.31 centigrade. From the ice frazil forming procedure, when the water temperature decrease below freezing point, the ice frazil start to generate and its density increase following the heat exchange, and the water temperature variation tends to stability.

Figure 4.11 shows the calculated water temperature and ice frazil at every cross-sections before the river freeze-up, because the location of the first three cross-sections very close to the input water body, which the distance of YC 70 to input is 1,600 m, and with the influence of input temperature 0.0 centigrade, the water temperature of the first three cross-sections remain from -0.05 to 0.0 centigrade, and a small amount ice frazil generated. From the YC 74, the water temperature decrease to below -0.3 centigrade, and with a large number of ice frazil generated. Until calculated to 600 h (25 days), from YC 71 to YC 77 which all frozen-up. The Figure 4.12 shows the water temperature and ice frazil distribution after river reach freeze-up, due to the location of the first three cross-sections very close to the input water body, the ice frazil generated a little and cannot form the ice cover, from YC 71, especially form YC 75, with the influence of ice cover, the heat exchange value of water body with ice cover is smaller than that of water body with atmosphere, so the water temperature of the water body increases gradually, and ice frazil decreases gradually.

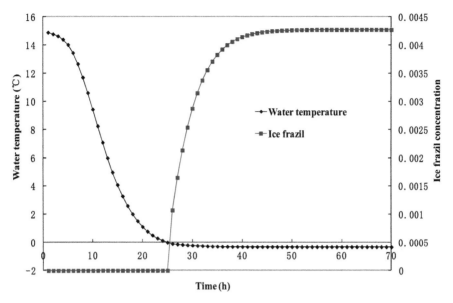

Figure 4.10 Water temperature and ice frazil variation at YC 77

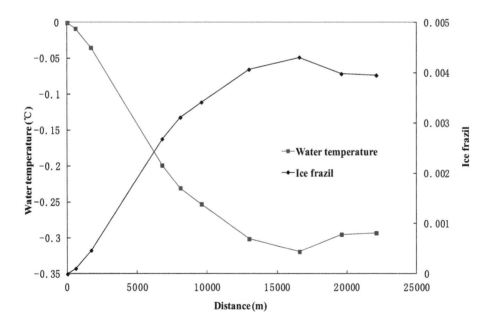

Figure 4.11 Calculated water temperature variation and ice frazil distribution (70h)

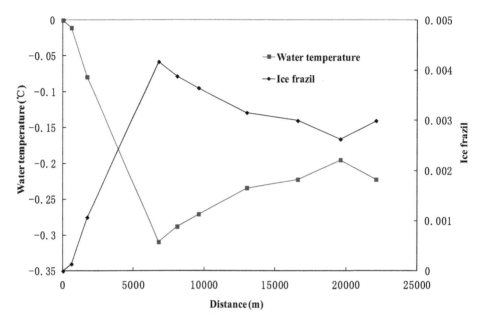

Figure 4.12 Calculated water temperature variation and ice frazil distribution (600h)

Water surface ice run concentration

The water surface ice run concentration is mainly influenced by the upstream water and ice, suspended ice frazil come-up and heat exchange between water body and atmosphere. Figure 4.13 shows the calculated ice run concentration variation at YC 70, YC 73, and YC 77 respectively. With the influence of initial river water temperature, and input water body temperature and outside meteorological conditions, the water temperature of YC 70 firstly decrease below freezing point and water surface ice floats generated, however, due to it is very close to the input, the water surface ice run concentration remains low, and during the open water period, which keeps at 0.036 basically. The water temperature of YC 73 decrease below freezing point at 20 h, in the open water period the water surface ice run concentration keeps at about 1.237. YC 77 as the output cross-section, mainly affected by the upstream incoming water and ice, the ice run concentration reach 0.248 and together with its boundary conditions and hydrodynamics conditions, which satisfied with condition to generate the ice bridge, then this cross-section firstly frozen, and with the development of the ice cover, its water surface ice run transform into ice cover, then the ice run concentration decrease and when this section frozen completely, the water surface ice run concentration tends to 0.

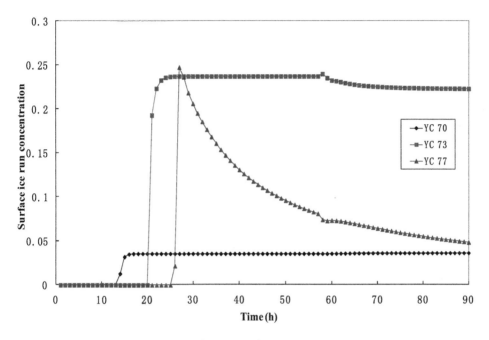

Figure 4.13 Calculated water surface ice run density variation

Land ice

Michel etc. (1982) proposed the empirical formula for land ice growth has applied conditions which is when $N >= 0.1$ and $0.167 < V_* < 1.0$ (N is surface ice concentration; $V_* = u/V_c$, u is velocity at depth of open water in front of the ice block, V_c is the maximum velocity which the surface ice can adhere to the land ice, takes 1.0 m/s), otherwise the land ice would break or cannot growth. Table 4.9 shows the land ice growth calculation results, without considering river freeze-up, when calculate to 600 h, according to the boundary conditions, hydrodynamics and thermodynamics conditions of the research reach, the YC 72 and YC 75 satisfied with the conditions of land ice growth, and land ice generate 0.19 m and 0.46 m respectively. Which means at that boundary and meteorological conditions, the land ice develops gradually.

Ice cover thickness

Table 4.10 shows the ice cover attributes at each cross-section. And Figure 4.14 show the ice cover thickness growth process, here YC 68-YC 77 represents the locations of the cross-sections as defined in Figure 3.4. YC 75 has satisfied with the river freeze-up conditions at 28 h, and YC 76 and YC 77 frozen at 29 h, thus these three cross-sections first frozen and ice cover formed. With the influence of the frozen of downstream, the backwater formed in front of the ice cover, and changed the hydrodynamics conditions, with the generation of the ice frazil, the cross-sections which located upstream satisfied with the river freeze-up conditions gradually, then from YC 71 to YC 74, the river freeze-up occurred at 292 h, 261 h, 259 h, and 160 h respectively. And for the cross-sections from YC 68 to YC 70, sue to the low ice frazil concentration, after the end of calculation, there is no river freeze-up. After river freeze-up, with the influence of meteorological conditions, ice cover thickness increase gradually, until the end of calculation, the output YC 77, the ice cover thickness reached 0.57 m, and YC 71 reached 0.28 m, and had the tendency of increasing continuously.

Table 4.9 Land ice growth calculation results at 600 h

Cross-section	$N = \dfrac{Q_i^s}{ut_i B}$	V_*	Land ice or not	River water surface width B (m)	River water surface width except land ice B_0 (m)	Land ice (m)
YC 68	0.0000	1.5356	No	327.66	327.66	
YC 69	0.0041	1.7894	No	277.34	277.34	
YC 70	0.0357	0.4780	No	809.62	809.62	
YC 71	0.0990	1.3122	No	619.35	619.35	
YC 72	0.1342	1.5466	Yes	446.19	446.00	0.19
YC 73	0.2374	2.2317	No	189.44	189.44	
YC 74	0.1602	1.7173	No	425.55	425.55	
YC 75	0.2892	1.4494	Yes	329.05	328.60	0.45
YC 76	0.2196	1.8364	No	380.78	380.78	
YC 77	0.3269	1.7193	No	294.70	294.70	

If the velocity of cross-section lower than the critical velocity of ice frazil deposition, the transported ice frazil would deposit under the ice cover, and formed the ice frazil aggregate, like suspended ice blocks, if the velocity increase, the ice frazil aggregate would erosion. Figure 4.15 shows the ice frazil aggregate process at each cross-section. After the ice cover formed, there were ice frazil aggregated, at YC 77, the ice frazil thickness reached 0.54 m, and had the tendency of increasing gradually.

Table 4.10 Ice cover attributes at each cross-section

Item	YC 68	YC 69	YC 70	YC 71	YC 72	YC 73	YC 74	YC 75	YC 76	YC 77
Initial freeze-up time (h)	-	-	-	292	261	259	160	28	29	29
Ice cover thickness (m)	-	-	-	0.28	0.32	0.35	0.41	0.51	0.59	0.57

Note: "-" means unfreeze-up.

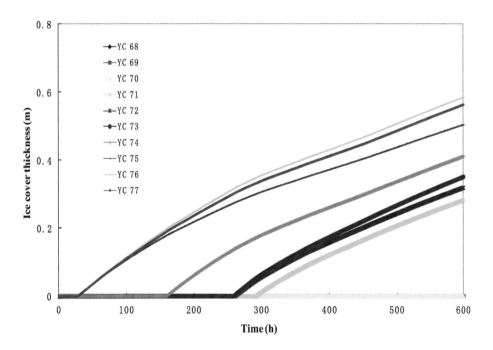

Figure 4.14 Ice cover thickness growth process

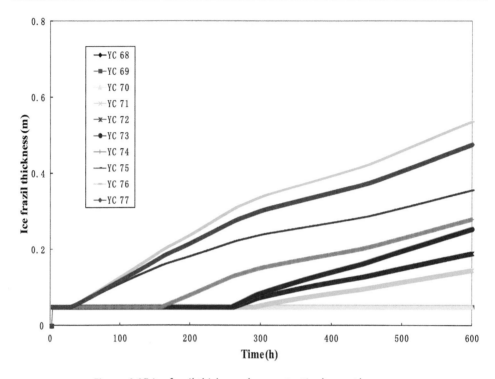

Figure 4.15 Ice frazil thickness (concentration) growth process

Ice cover roughness

Figure 4.16 shows the calculated ice cover roughness and composite roughness slope variation at YC 77. After the ice cover initial formed, the ice cover roughness is about 0.015, with the increase of the ice cover thickness, the roughness increase to 0.03 gradually, following the ice cover growth, the roughness has the tendency to decrease gradually. From Figure 4.16, at the beginning of the ice cover formed, the composite roughness slope is 1.84 times of that of river course roughness slope, following with the ice cover thickness increase, the composite roughness slope increase up to 4.77 times of that of river course roughness slope.

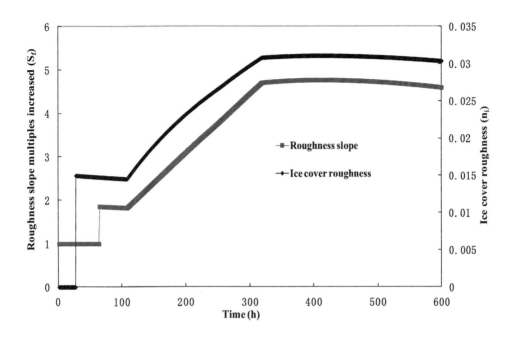

Figure 4.16 Ice cover roughness and composite roughness slope variation

Water level

Figure 4.17 shows the calculated water level variations under open water, freeze-up and ice covered period along the river course at 600 h. Without considering the influence of river ice, the water level comparison between open water and freeze-up shows the water level increase remarkably after the river freeze-up. Especially at YC.73, the maximum water level increase 1.25 m, which is the water level under the ice cover plus the 0.917 times the ice cover thickness, and the calculated water level under the ice cover is 0.54 m higher than that of open water period. Which is very similar with the observed water level increase about 1.4 m with discharge of 600 m³/s at Sanhuhekou in 1994. Then the calculation can reflect the formation of ice cover due to the influence of heading up water level.

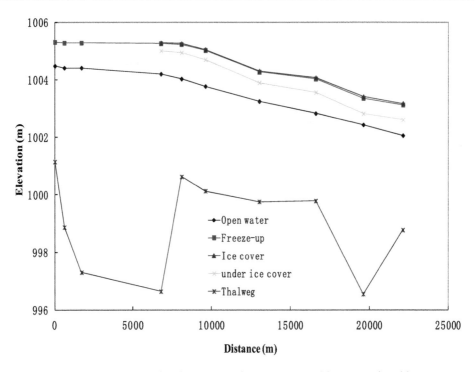

Figure 4.17 Water level variations for ice cover and freeze-up (600h)

4.6 Verification

Due to the available cross-section observation data and other hydrological and ice regime data, the reach (see Figure 3.4) from Sanhuhekou (YC 33) to Toudaoguai (YC 109) is selected for verification of the model. The river length is about 272 km as the research region, using the real time observed ice regime data and air temperature forecast data from air temperature forecast model to verify the one-dimensional ice flood numerical modelling, and the time series is winter period from 2013 to 2014. The observation items contains daily average discharge, daily average water level, daily average air temperature, daily average water temperature, and ice cover depth. The air temperature forecast data including the 1-10 days daily average air temperature at Sanhuhekou and Toudaoguai hydrometric stations.

4.6.1 Water temperature

The Figure 4.18 and Figure 4.19 shows the comparison of calculated with observed water temperature variation and observed air temperature at Sanhuhekou and Toudaoguai hydrometric station respectively. From November 1 with the air temperature decrease, the water temperature decrease gradually. For Sanhuhekou hydrometric station, the calculated water temperature reached freezing point on December 7, and the observed daily average water temperature decrease to freezing point on December 8, the difference is 1 day forward. And then the ice run occurred in the channel. In the ice run period, the calculated water temperature variation tendency is basically same with the observed. In the freeze-up period, the calculated water temperature is about freezing point which is consistent with the real conditions. In the river breakup period, the observed water temperature increase following with the increase of air temperature after March 23, the calculated water temperature increase on March 22, the difference is 1 day forward between calculated and observed.

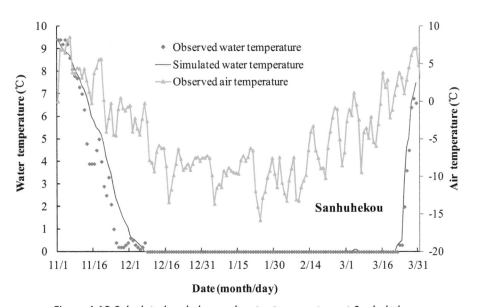

Figure 4.18 Calculated and observed water temperature at Sanhuhekou

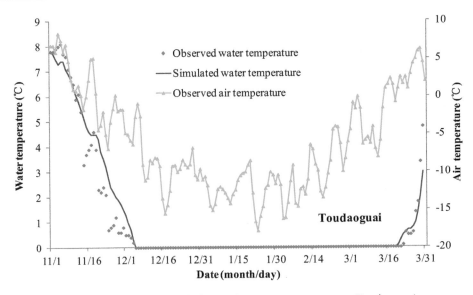

Figure 4.19 Calculated and observed water temperature at Toudaoguai

For Toudaoguai hydrometric station, similar with Sanhuhekou hydrometric station, the calculated water temperature and the observed daily average water temperature reached freezing point on December 5, and then the ice run occurred in the channel. In the river breakup period, the observed water temperature increase following with the increase of air temperature after March 22, the calculated water temperature increase on March 20, the difference is about 2 days forward between calculated and observed.

4.6.2 Ice run concentration

Figure 4.20 and Figure 4.21 show the comparison of calculated with observed ice run concentration variation and observed air temperature at Sanhuhekou and Toudaoguai hydrometric station respectively. For Sanhuhekou hydrometric station, the observed daily average water temperature decrease to freezing point on December 8, and the ice run occurred in the channel on December 5, which is 3 days earlier than the date when the water temperature decrease to freezing point, and the calculated one is on December 4, which is 1 day forward compare with the observed one. And then the ice run concentration increasing with the air temperature decrease, and on December 15, the observed ice run concentration reach maximum and means the river freeze-up, and the calculated one is on December 17, which is 2 days later. In the river breakup period, the observed ice run concentration decrease sharply after March 21 following with the increase of air temperature, and the calculated one is on March 20, which is 1 day forward between calculated and observed.

For Toudaoguai hydrometric station, the calculated ice run occurred in the channel on December 5, which is 1 day later than the date when the observed ice run occurred in the channel. The calculated ice run concentration reach the maximum one on December 17, which is 2 days later than the date when the observed ice run concentration reach the maximum, which mean the river freeze-up. In the freeze-up period, the calculated ice run concentration is consistent with the real conditions. In the river breakup period, the observed ice run concentration decrease gradually following with the increase of air temperature, the calculated ice run concentration decrease after March 17, the difference is about 2 days forward between calculated and observed values.

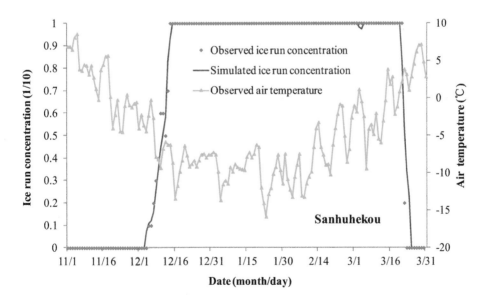

Figure 4.20 Calculated and observed ice run concentration at Sanhuhekou

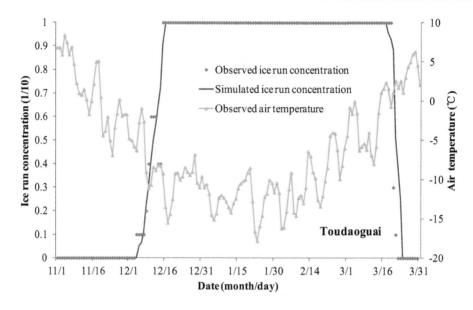

Figure 4.21 Calculated and observed ice run concentration at Toudaoguai

4.6.3 Ice cover thickness

Figure 4.22 and Figure 4.23 shows the comparison of calculated with observed ice cover thickness variation and observed air temperature at Sanhuhekou and Toudaoguai hydrometric station respectively. For Sanhuhekou hydrometric station, the observed initial freeze-up time is on December 14 and the calculated is on December 6, the calculated one is 8 days earlier than the observed one. The calculated maximum ice cover depth is 52.1 cm on February 2 and the observed one is 50.0 cm from January 21 to February 6, and the difference is about 2.1 cm more and 4 days later. During the period from February 6 to March 1, although the daily average air temperature was below freezing point, the daily air temperature has the increase tendency, and the ice cover depth has the tendency of decrease, the calculated has the same variation tendency with the observed. After March 1, with the influence of air temperature increasing rapidly, the ice cover depth decrease sharply, the calculated ice cover thickness decrease to 0 m is on March 8 and the observed one is on March 6, the calculated ice cover disappear date is 2 days later than that of the observed.

For Toudaoguai hydrometric station, the observed initial freeze-up time is on December 15 and the calculated is on December 5, the calculated one is 10 days earlier than the observed one. After the ice cover formed, with the effect of heat exchange between ice and air and between ice and water, the ice cover depth increase gradually, the tendency of ice cover depth variation is basically same. The calculated maximum ice cover depth is 75.0 cm on February 16 and the observed one is 74.0 cm from January 15 to February 16, and the difference is about 1 cm more. During the period from February 16 to March 15, although the daily average air temperature was below freezing point, the daily air temperature has the increase tendency, and the ice cover depth has the tendency of decrease, the calculated has the same variation tendency with the observed. After March 15, with the influence of air temperature increasing rapidly, the ice cover depth decrease sharply, the calculated ice cover thickness decrease to 0 m is on March 25 and the observed one is on March 19, the calculated ice cover disappear date is 6 days later than that of the observed.

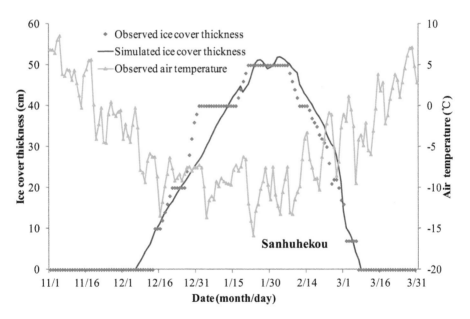

Figure 4.22 Calculated and observed ice cover thickness at Sanhuhekou

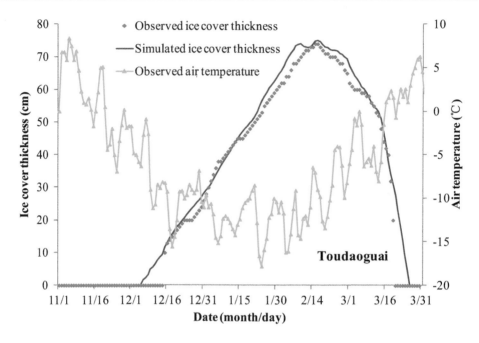

Figure 4.23 Calculated and observed ice cover thickness at Toudaoguai

4.6.4 River freeze-up and breakup date

The observed river freeze-up date for Sanhuhekou and Toudaoguai is on December 14 and December 15 respectively. Based on the actual conditions of the Ning-Meng reach, when $\delta \geq 0.012$ or the ice run concentration C_i is larger than 75%, the river will freeze-up, then using the river freeze-up judgment together with the forecasted air temperature and observed discharge data to calculate the river freeze-up date is on December 15 and December 16 with the lead-time of 5 to 6 days respectively, and the difference between observation and simulation are all 1 day later.

The observed river breakup date for Sanhuhekou and Toudaoguai is on March 22 and March 20 respectively. Using the river breakup judgment together with the forecasted air temperature and observed ice cover thickness data to calculate the river breakup date is on March 21 and March 19 with the lead-time of 5 to 6 days respectively, and the difference between observation and simulation are all 1 days earlier.

The river freeze-up and breakup date with 5 to 6 days lead-time is fundamental information used for decision making of ice flood control, to decide when and how to control the projects such as Liujiaxia reservoir upstream and Wanjiazhai reservoir downstream, to let the river thermal freeze-up and breakup, avoid the occurrence of the river mechanical freeze-up and breakup.

4.6.5 Discharge

In order to analyse the influence of frozen days on the water balance of a river reach, a comparison of the simulated discharge variation in two situations is done for Sanhuhekou and Toudaoguai hydrometric stations. The two simulations are done as follows: one in which the number of freezing days and ice formation are taken into consideration; and a second one in which freezing is not taken into consideration. Results are presented in Figure 4.24 and Figure 4.25 for hydrometric stations Sanhuhekou and Toudaoguai, respectively. The tendency of the simulated discharge with considering the influence of frozen and water balance on the water body and the observed one is almost similar. For the simulated discharge without considering the influence of frozen and water balance on the water body, during the river freeze-up period is higher than the observed one, and during the river breakup period, the appearance of discharge is on the contrary which is lower than the observed one. The reason is that during the river freeze-up period, the water transforms into ice, which decreases channel water storage and causes water loss, while during the river breakup period, water releases from channel water storage causing water volume to increase.

For Sanhuhekou hydrometric station, after the river freeze-up, the simulated discharge decrease sharply, the reason is most of the water transform into ice, which is channel water storage, and the observed minimum discharge is 250 m^3/s on December 20, and the calculated one is 220 m^3/s on December 27, the difference is 30 m^3/s less and 7 days later. During the river breakup period, the observed maximum discharge, which is ice flood peak, is 810 m^3/s on March 23, and the calculated one is 780 m^3/s on March 23, the difference being 30 m^3/s less.

For Toudaoguai hydrometric station, after the river freeze-up, the discharge decrease sharply, and the observed minimum discharge is 126 m³/s on December 16, and the calculated one is 149 m³/s on December 17, the difference is 23 m³/s more and 1 days later. During the river breakup period, the observed maximum discharge which is ice flood peak is 1450 m³/s on March 27, and the calculated one is 1535 m³/s on March 28, the difference is 85 m³/s more and 1 days later.

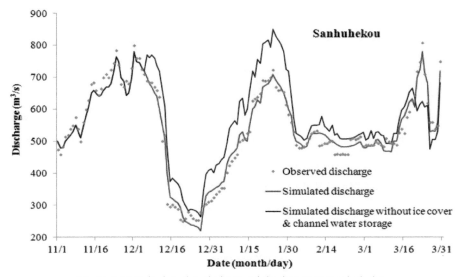

Figure 4.24 Calculated and observed discharge at Sanhuhekou

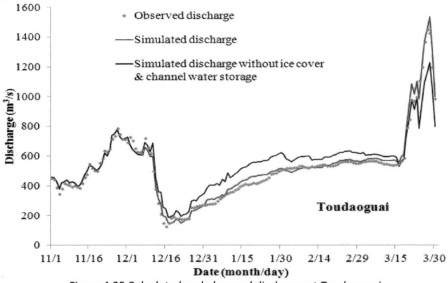

Figure 4.25 Calculated and observed discharge at Toudaoguai

118

4.6.6 Water level

In order to analyse the influence of frozen on the water body, comparison of the water level variation counting frozen and without counting frozen at Baotou hydrometric station (Figure 4.26) is used. After December 6 with the influence of frozen, the friction energy dissipation generated by ice cover on the water body, and the water body need to increase potential energy, and the calculated water level will increase. The ice cover depth and frozen time which all impact the friction of ice cover. The simulated water level considering the influence of ice cover is higher than that without considering the influence of ice cover, which means with the influence of ice cover, the flow friction increase and impel the water level to increase corresponding. The calculated ice cover friction reached the maximum value on December 23, and the corresponding water level is 1,002.85 m. After December 23, the ice cover frozen further, and the ice cover roughness coefficient decrease following the water level decrease gradually. Although after March 22, the cross-section has been breakup, with the release of water channel-storage, the calculated water level is higher than that without counting ice cover. The observed maximum water level is 1003.88 m and on March 23, and the calculated one is 1003.99 and on March 25, the difference is 0.11 m more and 2 days later.

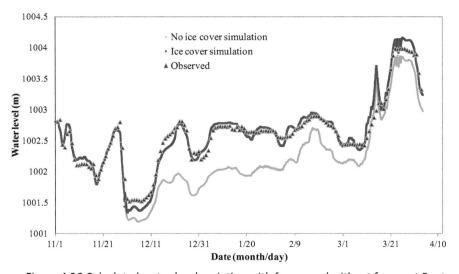

Figure 4.26 Calculated water level variation with frozen and without frozen at Baotou

During the ice cover breakup, the channel water storage will concentrated release, which result in the water volume increase in the river course, together with the effect of the ice blocks which increase the jamming, the water level reached the highest ever.

4.6.7 Channel water storage

Channel water storage cannot measure directly, according to the water balance principle, using the upstream and downstream observed discharge from hydrometric cross-sections to calculate the channel water storage. From the initial river freeze-up to start calculation, according to the upstream flow, and the observed discharge data, considering the factors such as flow propagation time and water level variation, to calculate the channel water storage for each river reach. The velocity and flow propagation times for different discharge are shown in Table 4.11 and Table 4.12.

Table 4.11 Statistical velocity values for different discharge at hydrometric stations

Discharge (m^3/s)	Lanzhou (m/s)	Shizuishan (m/s)		Bayangaole (m/s)		Sanhuhekou (m/s)		Toudaoguai (m/s)	
	Open water	Open water	Freeze-up	Open water	Freeze-up	Open water	Freeze-up	Open water	Freeze-up
200	1.00	0.89	0.26	0.73	0.47	0.63	0.41	0.52	0.41
300	1.10	0.98	0.39	0.87	0.55	0.76	0.49	0.61	0.46
400	1.20	1.07	0.52	1.01	0.63	0.89	0.57	0.70	0.51
500	1.30	1.16	0.65	1.15	0.71	1.02	0.65	0.79	0.56
600	1.40	1.25	0.78	1.29	0.79	1.15	0.73	0.88	0.61
800	1.60	1.43	1.04	1.57	0.95	1.41	0.89	1.06	0.71
1000	1.80	1.61		1.85		1.67		1.24	
1200	2.00	1.79		2.13		1.93		1.42	
1500	2.30	2.06		2.55		2.32		1.69	

Table 4.12 Statistical flow propagation times for different discharge at hydrometric stations

Discharge (m³/s)	Liujiaxia to Lanzhou (d) Open water	Shizuishan (d) Open water	Bayangaole (d) Open water	Bayangaole (d) Freeze-up	Sanhuhekou (d) Open water	Sanhuhekou (d) Freeze-up	Toudaoguai (d) Open water	Toudaoguai (d) Freeze-up
200	1	7	9	10	12	14	17	22
300	1	6.5	8.5	9.5	11	13.5	16	20
400	1	6	7	8	9	11	14	17
500	1	6	7	8	9	11	14	17
600	1	6	7	8	9	11	14	17
800	0.5	5.5	6.5	7.5	8.5	10.5	11	14
1000	0.5	5.5	6.5		8.5		11	
1200	0.5	5	6		7.5		10	
1500	0.5	5	6		7.5		10	
Distance (km)	99	581	142		221		300	

The method to calculate the discharge variation, i.e. the Muskingum routing method, was used to calculate the discharge variation, which is a simple hydraulic method. However, the shortage of this method is that it cannot provide channel water storage spatial distribution and cannot divide the total water storage as main channel storage and flood plain storage.

Figure 4.27 shows the channel storage increment variation from Sanhuhekou to Toudaoguai. The variation tendency of the calculated with the observed is almost same. The calculated maximum channel storage is 1.023 billion m³, and the observed is 1.095 billion m³, and the calculated is less than the observed with 0.072 billion m³, and the occurrence date of the calculated with the observed is 3 days late. From the source of the channel storage, the main channel ice cover storage has reached the maximum value of 0.118 billion m³ as well as the ice cover depth reached the maximum one. In the beginning of the freeze-up period, the ice cover of the main channel is the key component of the channel storage, and during the stable freeze-up period, the variation of the channel storage mainly influenced by the upstream water amount, and reached the maximum value of 0.276 billion m^3 on February 23. According to the channel conditions, the flood plain also the main place for water storage. Around December 15, the calculated the flood plain started to store water, following the ice cover increase and the continuous upstream water, the flood plain water storage increase gradually, during the stable freeze-up period, with the relatively stable input and output discharge, and less ice cover roughness variation, the flood plain water storage remain stable, and the maximum one reached 0.649 billion m^3 on February 23. When the air temperature increase above freezing point, the channel storage would melt and moves back to the river channel. So based on the above analysis, the flood plain ice cover and its storage is the main source of the channel storage increment, which takes 63.44% of the maximum channel storage increment, and then the main channel storage increment, which takes 26.98% of the maximum one, and the main channel ice cover storage is relatively stable and takes about 10.0% of the maximum one.

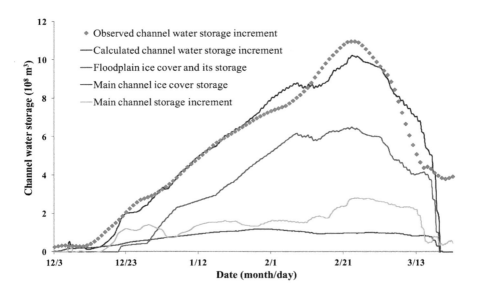

Figure 4.27 Channel water storage variation from Sanhuhekou to Toudaoguai

When the river freeze-up, with the effect of ice cover friction, the flow capacity under ice cover will decrease, and the discharge to the downstream will decrease, some part of the water volume will stay in the channel and formed the channel storage, and increase the upstream water level, and change the open flow relation between the water level and discharge. Generally, more ice cover friction, less flow capacity under the ice cover, result in more channel water storage and high water level.

4.7 Analysis and discussion

Tracking ice formation from observations and combining them with numerical model predictions for advanced warning requires proper understanding of all scientific issues that play a role. However, it is not possible to make specific predictions because our physical understanding remains incomplete, thus the main challenge is how to accelerate the pace of discovery and bridge the major knowledge gaps. In the case of the Yellow River, ice floods impose a threat every year, which is why the YRCC is putting considerable effort in verifying theoretical formulations with actual field measurements in order to better understand the scientific mechanisms that play a role.

In the case of the Yellow River, the parameters for ice cover roughness, decay constant, heat exchange coefficient between water and atmosphere, water temperature calculation, and the judgement regarding river freeze-up and breakup have been determined using observed hydrometeorological data. It was especially noted that for the roughness of ice cover and the decay parameter, considering the ice cover friction varies during the ice cover formation according to different ice regime conditions.

In order to test the results of ice flood model, based on the available cross-section observation data and other hydrological and ice regime data, selecting the reach from Zhaojunfen to Huajiangyingzi as the case test reach for the ice flood numerical modelling, to design the ice regime, meteorological, boundary, initial input and output etc., conditions, to simulate the generation and disappearance procedure of river ice. The simulation and calculation of river ice procedures including water temperature variation and frazil ice generation procedure, water surface ice run concentration, land ice, ice cover thickness, ice cover roughness, water level. The results analysis shows the mathematical representations of the ice flood modelling for the Ning-Meng reach are rational, and outputs are reasonable and acceptable. As ice induced flood data collection continues in different regions around the world, the available database to determine the possible times of ice formation is enlarged and improved (Debolskaya, 2009). Based on ice formation predictions decision makers can take appropriate measures to reduce the risk of flooding. Flooding during cold season is very important, therefore, determination of the moments of ice formation that could possibly eliminate flooding, due to the decisions taken is also an important task in modelling. Selecting the river reach from Sanhuhekou to Toudaoguai as typical reach to verify the coupling modelling of hydrodynamics and thermodynamics, which using the real time observed ice regime data, meteorological data to validate and verify the ice numerical modelling for the research and simulation of river ice phenomena.

Table 4.13 Calculated and observed ice regime at Sanhuhekou and Toudaoguai

Item	Sanhuhekou			Toudaoguai		
	Observed	Simulated	Difference	Observed	Simulated	Difference
Water temperature to freezing point date	Dec.8	Dec.7	1 d	Dec.5	Dec.5	0 d
Water temperature above freezing point date	March 23	March 22	1 d	March 22	March 20	2 d
Initial ice run date	Dec.5	Dec.4	1 d	Dec.5	Dec.4	1 d
Ice run end date	March 23	March 25	2 d	March 19	March 17	2 d
Maximum ice cover thickness	50.0 cm	52.1 cm	2.1 cm	74 cm	75 cm	1 cm
Maximum ice cover thickness date	Feb.2	Feb.6	4 d	Feb.16	Feb.16	0 d
Initial freeze-up date	Dec.14	Dec.15	1 d	Dec.15	Dec.16	1 d
Breakup date	March 22	March 21	1 d	March 20	March 19	1 d
Minimum discharge	250 m^3/s	220 m^3/s	30 m^3/s	126 m^3/s	149 m^3/s	23 m^3/s
Minimum discharge date	Dec.20	Dec.27	7 d	Dec.16	Dec.17	1 d
Flood peak	810 m^3/s	780 m^3/s	30 m^3/s	1450 m^3/s	1535 m^3/s	85 m^3/s
Flood peak date	March 23	March 23	0 d	March 27	March 28	1 d

Table 4.13 shows the calculated and the observed ice regime items at Sanhuhekou and Toudaoguai. Though from the analysis and comparison mentioned above, we can get the conclusion that for the water temperature, ice run concentration, ice cover thickness, river freeze-up and breakup date, discharge, water level and channel water storage, the ice flood model simulation results are acceptable and reasonable. The ice flood numerical modelling of the Ning-Meng reach is applicable to the Ning-Meng reach for simulating ice regimes, and can be used to forecast the ice regime to support decision making, such as on artificial ice-breaking and reservoir regulation.

Especially using the forecasted air temperature data as input for running the model, this can prolong the lead-time and during the river freeze-up and breakup period, the ice regime can be predicted and support decision making, which can be used for consideration of reservoir regulation and other engineering measures, such as Liujiaxia reservoir regulation to control the river as thermal freeze-up and breakup, for diminish the possibility of ice jam and ice dam occurrence, Wanjiazhai reservoir regulation to make the artificial flood to flush the sediment at the Tongguan Heights.

Although the present research focused on ice formation rather than on floods, it can be generally concluded that the measured elements and frequency should be increased, and as recommendation if floods need to be captured and simulated, then one-dimensional models should be extended to two-dimensional models. This will allow to account for better water levels and volumes of water that are outside of the river bed, where ice formation would be for longer period of times. The simulation of spatial and temporal distribution of the channel water storage will be the most important account for more extreme ice floods.

Chapter 5   Sensitivity and Uncertainty Analysis

5.1 Sensitivity analysis

The One-At-A-Time sensitivity method was used to conduct the sensitivity analysis. That means changing the value of one parameter from the minimum value to the maximum value, and at the same time keeping other parameters constant at their mean value, then check the variation of the model output. The data used for performing sensitivity analysis includes cross-sections and bed elevation, discharge, water level, ice cover thickness, air temperature, and water temperature at Bayangaole, Sanhuhekou, and Toudaoguai hydrometric station, during the winter of 2007/2008 and 2008/2009, which were two quite severe winters, much more so than the years 2012/2013/2014 that were used in the previous chapter. The sensitivity parameters are presented in Table 5.1 and the 6 cases were designed to check the parameter sensitivities as shown in Table 5.2. The result of the 6 cases designed to perform sensitivity analysis are shown in this section; firstly, the sensitivity of the parameters is discussed to prepare for the calibration of the model; secondly, the result of sensitivity analysis at each case is analyzed in order to check the accuracy of the model qualitatively.

Table 5.1 Sensitivity of parameters in the river ice flood model

Sensitivity parameter	Physical meaning	Unit	Reference range
n_b	Bed roughness	-	0.019-0.045
$n_{i,e}$	End ice roughness	-	0.008-0.035
α_n	Decay constant	1/day	0.005-0.05
Coe_Cw	Heat exchange coefficient between water and ice	-	15-18
Coe_Hia	Heat exchange coefficient between ice and atmosphere	-	6-12
Coe_Co	Heat exchange coefficient between water and atmosphere	-	15-20

Table 5.2 Cases used for sensitivity analysis

Item	α_n	$n_{i,e}$	n_b	Coe_Cw	Coe_Hia	Coe_Co
Case 1-a	0.005	0.0215	0.032	16.5	9	20
Case 1-b	0.05	0.0215	0.032	16.5	9	20
Case 2-a	0.2025	0.008	0.032	16.5	9	20
Case 2-b	0.2025	0.035	0.032	16.5	9	20
Case 3-a	0.2025	0.0215	0.019	16.5	9	20
Case 3-b	0.2025	0.0215	0.045	16.5	9	20
Case 4-a	0.2025	0.0215	0.032	15.0	9	20
Case 4-b	0.2025	0.0215	0.032	18.0	9	20
Case 5-a	0.2025	0.0215	0.032	16.5	6	20
Case 5-b	0.2025	0.0215	0.032	16.5	12	20
Case 6-a	0.2025	0.0215	0.032	16.5	9	15
Case 6-b	0.2025	0.0215	0.032	16.5	9	20

Case 1   Decay values change from 0.005 to 0.05, others remain at mean values

From Figure 5.1 the following two conclusions can be drawn:

(1) The water level at Sanhuhekou station is sensitive to the decay constant value, but the others are not sensitive to the decay constant value.

(2) The output of the model seems as expected. The decay constant value is related to the Manning coefficient under ice cover during the river freeze-up period, hence the water level and discharge differs at the beginning of the river freeze-up period. In this case, if the decay constant value increases, it means that the Manning coefficient under the ice cover would decrease faster, and the roughness of ice cover would decrease, which results in a decrease of equivalent roughness, so finally the water level would decrease.

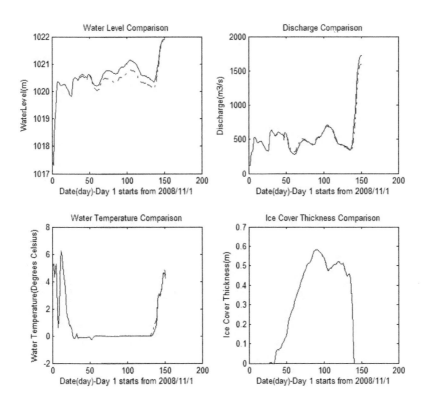

Figure 5.1 Sensitivity analysis of decay constant at Sanhuhekou
(Blue solid line: minimum value; Red dashed line: maximum value)

Case 2   End ice roughness changes from 0.008 to 0.035, others at mean values

From Figure 5.2, the following conclusions can be drawn:

(1) The water level at Sanhuhekou station is sensitive to the end ice roughness, but the others are not sensitive to it.

(2) The output of the model is acceptable. End ice roughness is related to the Manning coefficient under ice cover during the river freeze-up period, hence the water level and discharge differs at the beginning of the river freeze-up period. When the end ice roughness increases, it means that the Manning coefficient of the ice cover at the end of the river freeze-up period would increase, and then the equivalent roughness would increase, which could result in an increase in water level.

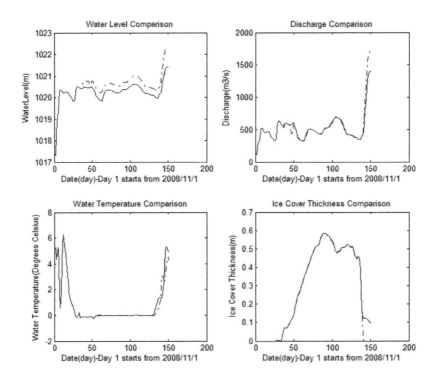

Figure 5.2 Sensitivity analysis of end ice roughness at Sanhuhekou
(Blue solid line: minimum value; Red dashed line: maximum value)

Case 3   Riverbed Manning coefficient changes from 0.019 to 0.045, others at mean values

From Figure 5.3, the following two conclusions can be drawn:

(1) The water level at Sanhuhekou station is sensitive to the Manning coefficient of river bed, but the others are not sensitive to the Manning coefficient of river bed.

(2) The output of the model seems as expected; Manning coefficients of riverbeds are related to the Manning coefficient during the whole period, hence the water level and discharge differs at the beginning of the simulation period. When the Manning coefficient of the riverbed increases, the equivalent roughness increases, which results in increase of water level.

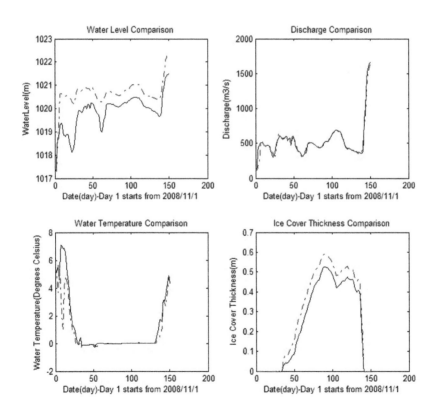

Figure 5.3 Sensitivity analysis of Manning coefficient of river bed at Sanhuhekou
(Blue solid line: minimum value; Red dashed line: maximum value)

131

Case 4   Water-Ice heat exchange coefficient variation

From Figure 5.4, it can be concluded that the water level, discharge, water temperature, and ice cover thickness at Sanhuhekou station are not sensitive to the heat exchange coefficient between water and ice.

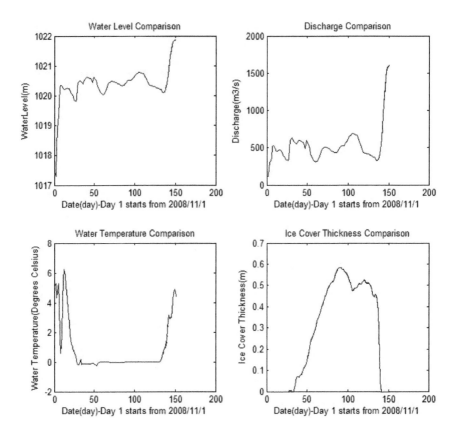

Figure 5.4 Sensitivity analysis of water-ice heat exchange coefficient at Sanhuhekou (Blue solid line: minimum value; Red dashed line: maximum value)

Case 5   Ice-Atmosphere heat exchange coefficient variation

From Figure 5.5, the following two conclusions can be drawn:

(1) The ice cover thickness at Sanhuhekou station is sensitive to heat exchange coefficient between ice and atmosphere, but water level, discharge, and water temperature not;

(2) The output of the model looks as expected: when the air temperature is lower than the ice temperature, the ice cover loses more heat, which results in increase of ice cover thickness. However, when the air temperature is higher than the ice temperature, the ice cover gains more heat, which results in a decrease of ice cover thickness. This is why the ice cover thickness becomes higher with the increase of heat exchange coefficient between ice and atmosphere from day 35 to day 125, and the ice cover thickness becomes lower with the increase of heat exchange coefficient between ice and atmosphere from day 125 to day 151.

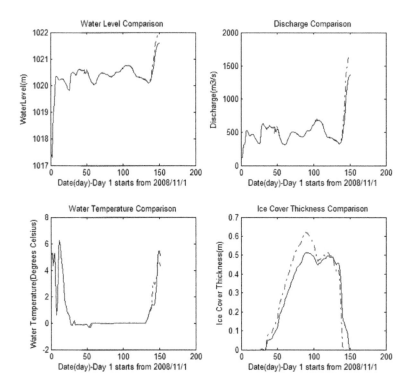

Figure 5.5 Sensitivity analysis of ice-atmosphere heat exchange coefficient at Sanhuhekou
(Blue solid line: minimum value; Red dashed line: maximum value)

Case 6   Water–atmosphere heat exchange coefficient variation

From Figure 5.6, the following two conclusions can be drawn:

(1) The ice cover thickness at Sanhuhekou station is sensitive to heat exchange coefficient between water and atmosphere, but the water level, discharge, and water temperature are not sensitive to the heat exchange coefficient between water and atmosphere.

(2) The output of the model is acceptable. Heat exchange coefficient between water and atmosphere is related to the heat exchange between water and atmosphere, if the heat exchange coefficient between water and atmosphere increased, it means that when the air temperature is higher than water temperature, the water could gain more heat, which could result in the increase of water temperature, and then the ice cover thickness could decrease, and vice versa.

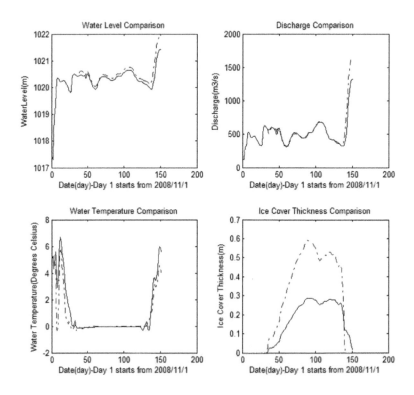

Figure 5.6 Sensitivity analysis of water and gas heat exchange coefficient at Sanhuhekou
(Blue solid line: minimum value; Red dashed line: maximum value)

The results of the sensitivity analysis are summarized in Table 5.3.

Table 5.3 Results of sensitivity analysis

Sensitivity parameters	Physical meaning	Reference range	Unit	Sensitive object
n_b	Bed roughness	0.019-0.045	-	Water level
$n_{i,e}$	End ice roughness	0.008-0.035	-	Water level
α_n	Decay constant	0.005-0.4	1/day	Water level
Coe_Cw	Heat exchange coefficient between water and ice	15-18	-	-
Coe_Hia	Heat exchange coefficient between ice and atmosphere	6-12	-	Ice cover thickness
Coe_Co	Heat exchange coefficient between water and atmosphere	15-20	-	Ice cover thickness

The One-At-A-Time sensitivity method was used to conduct the sensitivity analysis. The results show that bed roughness, end-ice roughness, and decay constant are sensitive to water level variations at Sanhuhekou station, and that the heat exchange coefficient between ice and air and between water and air are sensitive to ice cover thickness at the Sanhuhekou station.

5.2 Uncertainty analysis

Based on the sensitivity analysis result, the water level at Sanhuhekou station is sensitive to Manning coefficient of the riverbed, decay constant, and ice cover roughness. The ice cover thickness at Sanhuhekou station is sensitive to heat exchange coefficient between ice and atmosphere, and heat exchange coefficient between water and air. Hence, the uncertainty analysis is divided into uncertainty analysis about water level and ice cover thickness at Sanhuhekou station, and Monte Carlo simulation is used to conduct the parametric uncertainty analysis (Moya-Gomez et al., 2013).

The scenarios of the uncertainty analysis about water level at Sanhuhekou station are designed based on the calibrated parameters, the related three parameters which the parameters are tested and verified, the values can be seen specifically in Table 5.4, the range was calculated through increasing and decreasing the calibrated value by 20%, and the samples generation was uniformly random, and the number of simulations was 400. When conducting the uncertainty analysis about the ice cover thickness at Sanhuhekou station, the related two parameters are heat exchange coefficient between ice and air, and heat exchange coefficient between water and air. The scenarios of the uncertainty analysis about ice cover thickness at Sanhuhekou station are designed based on the calibrated parameters, which could be seen in Table 5.5, the range was calculated through increasing and decreasing the calibrated value by 20%, and the samples generation was uniformly random, and the number of simulations was 400. Based on the above cases design, input the parameters into the model, run the model and store the result, analyze the distribution and quantiles of the output, namely the PDF (Probability Density Function) at one time step and two bounds (5% and 95%). After running the model, the observed data, 5% percentile bound, 95% percentile bound, and the result of 400 cases are shown in Figure 5.7. The mean value (50%) can easily be inferred by visual inspection (Figure 5.7), and can be compared with the observed values.

Table 5.4 Cases of uncertainty analysis about water level

Variable	Physical meaning	Range	Samples generation	Unit
n_b	Bed roughness	0.012-0.020	Uniformly random	-
α_n	Decay constant	0.0008-0.0012	Uniformly random	1/day
$n_{i,e}$	End ice roughness	0.008-0.012	Uniformly random	-

Table 5.5 Cases of uncertainty analysis about ice cover thickness

Variable	Physical meaning	Range	Samples generation	Unit
Coe_Hia	Heat exchange coefficient between ice and atmosphere	9-15	Uniformly random	-
Coe_Co	Heat exchange coefficient between water and atmosphere	15-20	Uniformly random	-

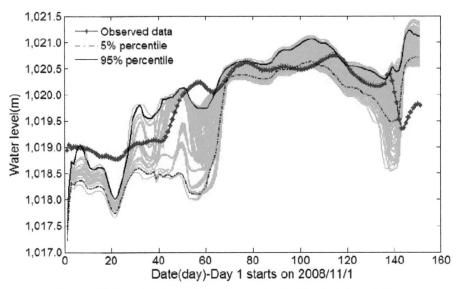

Figure 5.7 Uncertainty analysis of water level at Sanhuhekou station

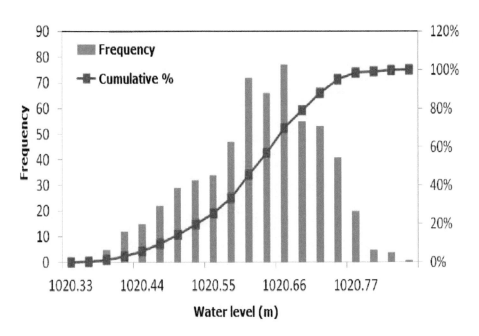

Figure 5.8 Probability distribution of water level on Day 90

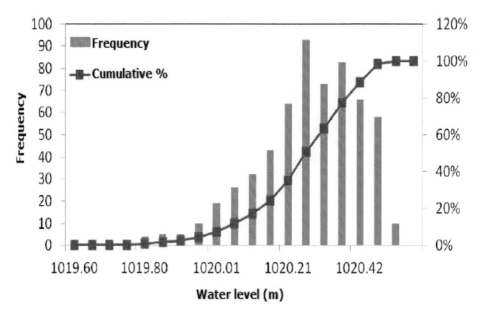

Figure 5.9 Probability distribution of water level on Day 130

Day 90 and Day 130 are chosen to show the probability distribution. When it comes to the probability distribution of uncertainty analysis result, if the distribution looks like normal distribution, the uncertainty analysis result is good. According to Figure 5.8 and Figure 5.9, the probability distribution of water level on day 90 looks good, that is because it looks like normal distribution. Although the probability distribution of water level on day 130 is skew normal distribution, at least it shows the basic shape of normal distribution, hence the uncertainty analysis result is acceptable.

The observed data, 5% percentile bound, 95% percentile bound, and the result of 400 cases are shown in Figure 5.10. Day 60 and Day 120 are chosen to show the probability distribution.

According to Figure 5.11 and Figure 5.12, the uncertainty analysis results are not so good, due to the fact that the probability distribution of the ice cover thickness on these two days is not yet a normal distribution. That is because the number of cases designed for the uncertainty analysis is only 400, which is not sufficient to show the full characteristics of the distribution.

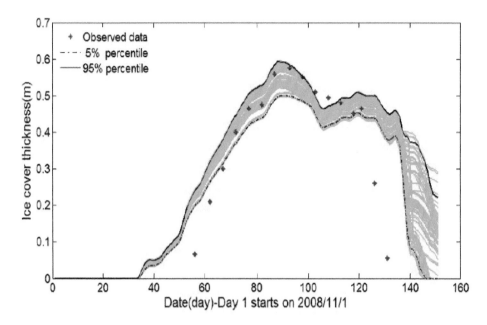

Figure 5.10 Uncertainty analysis about ice cover thickness at Sanhuhekou station

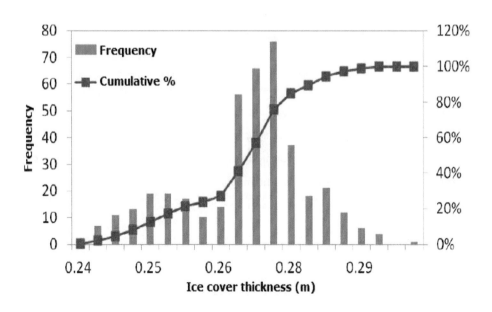

Figure 5.11 Probability distribution of ice cover thickness on Day 60

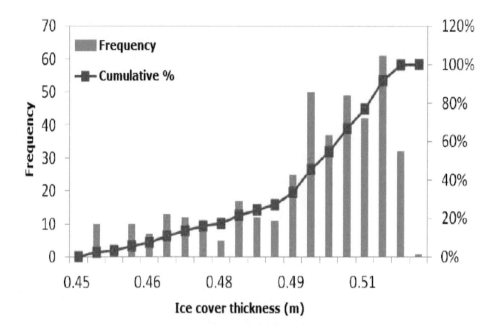

Figure 5.12 Probability distribution of ice cover thickness on Day 120

Based on the sensitivity analysis results, the uncertainty analysis is divided into (i) uncertainty analysis about water level and (ii) ice cover thickness at Sanhuhekou station resp. and a Monte Carlo simulation is used to conduct the parametric uncertainty analysis. The scenarios of the uncertainty analysis are designed based on the calibrated parameters, the range was calculated through increasing and decreasing the calibrated value by 20%, and the sample generation was uniformly random. Based on the above case design, the parameters were input into the model, the model was run and the results were stored, the distribution and quantiles of the output were analysed, namely the PDF at one time step and two bounds (5% and 95%).

5.3 Implications for ice flood control decision support system

An ice flood numerical model is developed as one of the essential components at YRCC. This model can be used to supplement the inadequacies in the field and lab studies which are being carried out to help understand the physical processes of river ice on the Yellow River. The main purposes of the ice flood numerical model are to satisfy the requirement of safeguard the ice flood and sufficiently utilize the limited water resources, the results of the model which are non-engineering measures and important information for ice flood control decision supporting. With the support of the ice flood numerical model, the decision makers can sufficiently regulate the reservoirs to control the base flow. Changing dam operation seems the best effective and economical way and can be realized easily.

For the ice flood control, the river freeze-up period and breakup period are most key periods which are easily to generate the ice jams and dams which cause the damages. During the river freeze-up period and breakup period, usually to cutoff the discharge advance through upstream reservoirs regulation, at the same time reduce the discharge of the reservoir will decrease the electricity power generation which should influence the human livings conditions and production of industry etc., therefore, in order to diminish the influence, the proper discharge should be determined according to the real ice regime conditions and channel conditions, and especially the most important is the date for river freeze-up and breakup should be foreseen as the supporting to reservoir regulations. For different dikes, there are different warning water level values, if the observed or forecasted water level is approaching the warning value, the upstream discharge will be reduced by reservoir regulation, or some water will be diverted to the ice flood retention area in case of emergency. For example, during the ice season from 2013 to 2014, the actual river freeze-up date is on December 12, the ice flood numerical model provide the river freeze-up date forecast is on December 11, with the lead-time of 6 days of 1 day difference, according to this information, the Ice Flood Control Headquarters Office of YRCC make the decision to reduce the discharge of Liujiaxia Reservoir and enlarge the discharge of Wanjiazhai reservoir. Furthermore, the actual river breakup date is on March 16 and 17 at Sanhuhekou and Toudaoguai hydrometric station respectively, the ice flood numerical model provide the river freeze-up date forecast is on March 17 and 18 December at Sanhuhekou and Toudaoguai hydrometric station respectively, with the lead-time of 6 days. according to this information, the Ice Flood Control Headquarters Office of YRCC make the decision to reduce the discharge of Longyangxia Reservoir properly and together with the forecasted flood peak and volume, to regulate the Wanjiazhai reservoir to form the artificial flood.

The channel water storage volume and its distribution which influence the river breakup ice flood peak and volume, these information is useful for decision making of the Wanjiazhai reservoir regulation to make the artificial flood procedure to flush and decrease the Tongguan Heights. For example, during the ice season from 2014 to 2015, regulate the Wanjiazhai reservoir with the peach ice flood volume information provided by the ice flood numerical model, to form the artificial flood to flush the Tongguan Heights, after that the elevation decrease 0.1 m, compare with other year's results, use less water volume (two thirds of the mean water volume from 2005 to 2010) to realize almost the same elevation decrease (the mean elevation decrease is 0.11 m from 2005 to 2010). Then the peach ice flood volume information provided by the ice flood numerical model helps to realize the fine regulation of the Wanjiazhai reservoir to flush the sediment at the Tongguan Heights.

Chapter 6   Conclusions and Recommendations

6.1 Conclusions

Based on several years of ice flood control experience in the Ning-Meng reach, it can be concluded that it is better to reduce the discharge in advance through upstream reservoir regulation, in order to decrease the opportunity for ice jams and ice dams to occur. For the Ning-Meng reach, the upstream reservoirs that can be used for ice flood control are (i) the Longyangxia reservoir and (ii) the Liujiaxia reservoir, which are located upstream at a distance of at least 779 km and a flow propagation time ranging from 7 to 23 days. This imposes a very high requirement on the ice regime forecast capabilities in both lead-time and precision. Following the development of an integrated water resources management plan for the Yellow River in 2006, YRCC is now utilizing the peach flood event to flush and decrease the Tongguan Heights, which requires accurate information such as when the flow peak occurs and what the peak release volume should be to regulate the Wanjiazhai reservoir by creating an artificial flood to flush the sediment from the Tongguan Heights.

Ice regime observation and forecasting is the most important scientific challenge for ice flood control, reservoir regulation and decision-making. Models are simplified representations of the real world (Mynett, 2002; Price, 2006). Therefore in order to solve the above mentioned problem, YRCC engaged in building a numerical ice flood model by coupling air temperature predictions from a medium range forecasting model for the Ning-Meng reach to a numerical ice flood model. In this way it proved possible to simulate and predict the ice regime processes during the ice flood period.

Hereafter each of the original research questions posed in Chapter 1 is summarized based on the results from the thesis, and recommendations towards further improvement are mentioned.

> *Specific field observations on ice regime can improve the representation and parameter selection of river ice processes and hydraulic modelling*

The formation, development and dissolution course of ice regime are mainly decided by the river pattern, the hydrometeorological regime and human activity. In each year, the ice regime is different because of the different hydrometeorological regime. Tracking ice formation from observations and combining with numerical model predictions for advanced warning requires proper understanding of all scientific issues that play a role. However, it is not possible to make specific predictions because our physical understanding remains incomplete, thus the main challenge is how to accelerate the pace of discovery and bridge the major knowledge gaps. In the case of the Yellow River, ice floods impose a threat every year, which is why the YRCC is putting considerable effort in verifying theoretical formulations with actual field measurements, such as setup more meteorological stations and add more ice regime characteristics and densify observations in order to better understand the scientific mechanisms that play a role. The accumulated in situ observation data can improve the understanding of representation and parameter selection of river ice processes for hydraulic modelling. Specific field observations on ice regime characteristics, especially the friction decay parameter and flow velocity distribution along the cross section, in combination with the analysis and understanding of the ice regime in the Ning-Meng reach, enable better possibilities to take measures to control and diminish ice floods hazards, and are useful for building and operating numerical ice flood models.

> *It is possible to investigate the effect of river ice cover friction on the river flow, and use the in situ observation data of actual ice regime conditions to quantify the ice cover roughness and establish the decay parameter for different hydrometeorological conditions*

During the river freeze-up period, the ice cover roughness plays an important role to influence the river flow, thus its influence cannot be neglected in the simulation of river ice flood modelling. If the river has a floating ice cover, these models take into account the roughness of the ice cover and consider the decay parameter as a constant; any lack of consideration of ice cover roughness varies during the ice cover formation and decay process, so the decay parameter should vary according to the different ice regime conditions. For the momentum equation of the Ning-Meng reach, we added an ice cover friction term to maintain the momentum balance. Since values for ice cover roughness vary during ice cover formation process, and decay parameter values vary according to different ice regime conditions, in situ observation data of actual ice regime conditions were used here to quantify the ice cover roughness and decay parameter at different hydrometeorological conditions.

It is possible to determine the changes in channel water storage capacity in case of ice formation, both in terms of spatial and temporal distribution as well as in growth and release processes

For the Ning-Meng reach of the Yellow River basin, where the river channel is flat and wide, the water level increases during the winter freeze-up period and a large amount of water flows into the floodplain to become a channel water storage compartment. However, after the water is frozen, it does not contribute to the water balance anymore. Inversely, during the breakup period of the river, the melting ice from the floodplains flows back into the main channel resulting in additional discharge. Therefore, for the ice flood modelling of the Ning-Meng reach, the channel water storage terms in the continuity equation include the ice cover term and the floodplain (melt) water discharge and ice term, to maintain the mass balance. The proposed channel water storage terms also provide a way to calculate the variations in channel water storage, which is of importance for supporting reservoir regulation during the river breakup period.

The river ice freeze-up and breakup criteria can be derived from combined thermodynamic, mechanical and hydrodynamic considerations.

As a result of the research in this thesis, new improved formulations are proposed for the Ning-Meng reach based on thermodynamic, mechanical and hydrodynamic considerations. Criteria are proposed in this thesis for river freeze-up and breakup time, using air temperature predictions, particular channel geometry, flow discharge and ice cover thickness, that indicate whether the river will freeze-up/breakup or not. Using the calibrated parameters together with the forecasted 1-10 days air temperature, proved very effective to identify the occurrence of river freeze-up and breakup with long lead-time. Based on this, the Ice Flood Control Headquarters Office of YRCC can make the decision to properly reduce the discharge of Longyangxia Reservoir and, together with the forecasted flood peak and volume, to regulate the Wanjiazhai reservoir to create an artificial flood for environmental reasons, if needed.

> *By coupling a numerical meteorological model with a numerical ice flood model, the lead time of ice regime forecasting and early warning can be prolonged.*

Meteorological numerical models are important tools for weather prediction and climate forecasting. The numerical ice flood model developed in this thesis critically depends on air temperature forecasts. By setting up a 1-10 days air temperature forecast model for Shizuishan, Linhe, Bayangaole, Sanhuhekou, Baotou, and Toudaoguai, we were able to obtain quite good results for the numerical ice flood model of the Ning-Meng reach. The coupling of a numerical meteorological model with a numerical ice flood model, using the air temperature forecast output of the meteorological model as input for the ice model, is the proper way to prolong the lead-time of ice regime forecasting and early warning.

> *A suitable and directly applicable numerical ice flood model was developed for the Ning-Meng reach of the Yellow River, with proper simulation of the ice regime including ice flood early warning and decision support.*

The numerical ice flood model for the Ning-Meng reach is applicable for simulating ice regimes, and has been used to forecast the ice regime and support decision-making, such as on artificial ice-breaking and reservoir regulation. Especially using the forecasted air temperature data as input for running the model can prolong the lead-time during the river freeze-up and breakup period. For different ice regime situations which can be predicted by the ice numerical model, there are different interventions or measures possible, such as modified reservoirs operation, diversion of the water to the retention area, bombing by airplane or artillery. It was found that by using the approach developed in this thesis, the ice regime can be predicted quite well to support decision making, which can be used for consideration of reservoir regulation and other engineering measures, such as regulating the Liujiaxia reservoir to control the river thermal freeze-up and breakup times, thus diminishing the possibility of ice jamming and the occurrence of ice dams, and for regulating the Wanjiazhai reservoir to make controlled artificial floods to flush e.g. the Tongguan Height. A suitable and directly applicable numerical ice flood model in the Ning-Meng reach of the Yellow River was developed, which considered the winter situations, by adding channel water storage terms in the continuity equation. Also an ice cover friction term was added to keep the momentum balance. Furthermore, empirical criteria were derived based on air temperature, channel geometry, discharge, and ice cover thickness, to judge whether river freeze-up or breakup may occur. Using these calibrated parameters together with 1-10 days forecasted air temperature proved very effective to predict river freeze-up and breakup with a longer lead-time for ice flood early warning and decision support.

6.2 Recommendations

A better understanding of river ice processes in combination with a better meteorological system could lead to further improvement of the ice modelling system. Modelling river ice processes, however, is difficult due to the complexities involved and interdisciplinary characteristics of the processes. Further research is needed on better understanding of thermal/mechanical processes in river ice modelling, especially the formation mechanism of ice jams and ice dams. Such kind of research needs more relevant domain knowledge, more spatial and temporal ice regime data and information, as well as in situ measurements and laboratory experiments. The collaboration between modellers and specific domain experts was one of the determining factors for the development of the river ice modelling in this research. It is very important to better the understanding of real world complex problems. Collaboration between modellers and domain experts need to be further strengthened in future research. It is very important to continue to take in situ measurements. No data-no results. Due to the complexity of the ice processes and the lack of instruments for measuring the ice regime, especially for the frazil slush density on the underside of ice covers and ice jam depth or ice dam depth, it is difficult for researchers to obtain that kind of data for analysis and getting to understand the mechanisms of the river ice processes. In the initial ice formation period, because of the thin ice depth, observers cannot take the measurements on the ice, and most of the time it is very dangerous for observers to measure ice jam or ice dam conditions. So there is a need to produce new and automatic instruments using sensors technologies to satisfy the requirement of the river ice processes research.

More sensitivity analyses can be done in future work for the ice floods numerical models, not only for individual processes or factors, but also for the interactions among different processes and factors. Such sensitivity analyses are very important for better analysing the interactions. Besides, because the key to understand complex systems often lies in understanding how processes and factors are related to each other at the hierarchical level, more sensitivity analyses need to be carry out including different factors as well as different processes in river ice floods modelling. Uncertainties in models and data are unavoidable, but can be reduced by further studies including better mathematical formulations and sensitivity analyses. More measurement data and better processing of available spatial and temporal data can further reduce uncertainty. Therefore, more in situ measurements need to be carried out to reveal the main processes and factors for river ice simulation. Still modelling is just one of the ways to obtain a better understanding of river ice phenomena and sometimes remains far from really understanding what is going on in river ice processes and ice transport processes. Therefore modelling results need to be augmented by combining specific domain knowledge from specialists with indoor laboratory experiments, as well as in situ observations, in order to have a better understanding of the real processes that take place.

Although the present research focused on ice formation rather than on floods, it can generally be concluded that the measurement quantity, quality and frequency should be increased. A recommendation for numerical flood simulation is to extend to 2D (preferably even 3D) models. Modelling applications need to be extended to include real (possible) situations, including the reservoir operations in the upstream part of the Yellow River and operating conditions of ice flood storage and detention along the Ning-Meng reach, to explore different scenarios. Using the numerical ice flood model to simulate these conditions, in order to provide more detailed information on river ice formation and transport processes, to facilitate decision-making for better ice flood control and water resources regulation management.

References

Aaltonen, J., Huokuna,M.,Severinkangas,K., and Talvensaari,M., 2008. Simulation of hanging dams downstream of Ossauskoski power plant. Proceedings of 19th International Association of Hydraulic Research Symposium on ice, Vancouver, British Columbia, Canada.Vol.1, pp.667-676.

Altberg, W.J., 1936. Twenty years of work in the domain of underwater ice formation: International Union of Geodesy and Geophysics, International Association of Scientific Hydrology, v.Bulletin 23, p.p. 373-407.

Ashton, G.D., 1979. Suppression of River Ice by Thermal Effluents. CRREL Report 79-30. Cold regions Research and Engineering Lab., U.S.Army, Hanover, NH. pp.23

Ashton, G.D.(Ed.), 1986. River and Lake Ice Engineering. Water Resources Publications, Littleton, Co.,U.S.A. pp.485.

Beltaos, S., 1983. River ice jams: theory, case studies and applications. Journal of Hydraulic Engineering , ASCE 109(10), pp.1338-1359.

Beltaos, S., 2008. Challenges and opportunities in the study of river ice processes. Proceedings of 19th International Association of Hydraulic Research Symposium on ice, Vancouver, British Columbia, Canada.Vol.1, pp.29-47.

Beltaos, S. and Krishnappan, B.G., 1982. Surges from ice jam release: a case study. Canadian Journal of Civil Engineering 9, pp.276-284.

Beltaos, S. and Wong, J., 1986. Downstream transition of river ice jams. Journal of Hydraulic Engineering, ASCE 112(2), pp.91-110.

Blackburn, J. and Hicks, F., 2003. Suitability of dynamic modeling for flood forecasting during ice jam release surge events. Journal of Cold Regions Engineering 17(1), pp.18-36.

Calkins, D.J., 1979. Accelerated Ice Growth in Rivers. CRREL Report 79-14, Hanover, NH, U.S.Army. pp.4.

Chen, D.L., Liu, J.F. and Zhang, L.N., 2012. Application of Statistical Forecast Models on Ice Conditions in the Ningxia-Inner Mongolia Reach of the Yellow River. Proceedings of 21th International Association of Hydraulic Research Symposium on ice, Dalian, China, pp.443-454.

Chen, S. and Ji, H., 2005. Fuzzy optimization neural network approach for ice forecast in the Inner Mongolia reach of the Yellow River. Hydrological Sciences 50(2), pp.319-330.

Chen, F. Shen, H.T., and Jayasundara, N., 2006. A One-Dimensional Comprehensive River Ice Model . Proceedings of 18th International Association of Hydraulic Research Symposium on ice, Sapporo, Japan.

Daly, S.F., 1984. Frazil ice dynamics. CRREL Monograph 84-1, U.S.Army CRREL, Hanover, N.H.

Daly,S.F. and Ettema, R., 2006. Frazil ice blockage of water intakes in the Great Lakes. Journal of Hydraulic Engineering , V. 132, p.p. 814-824, DOI:10.1061/(ASCE) 0733-9429 132:8 (814).

Debolskaya, E.I., 2009. Numerical Modeling of Ice Regime in rivers, in Hydrological Systems Modeling, Vol.I, pp.137-164, in Encyclopedia of Life Support Systems (EOLSS),Developed under the Auspices of UNESCO, Eolss Publishers, Osford, UK, [http://www.eolss.net].

Doering, J.C., Bekeris, L.E.,Morris, M.P., Dow, K.E. and Girling, W.C., 2001. Laboratory study of anchor ice growth: Journal of Cold Regions Engineering, V.15, pp. 60-66.

Fu, C., Popescu, I., Wang, C.Q., Mynett, A.E., and Zhang, F.X., 2014. Challenges in modelling river flow and ice regime on the Ningxia–Inner Mongolia reach of the Yellow River, China. Hydrology Earth System Science, Vol(18), doi:10.5194/hess-18-1225-2014, pp.1225-1237.

Gao, G.M., Yu, G.Q., Wang, Z.L. and Li, S.X., 2012. Advances in Breakup Date Forecasting Model Research in the Ningxia-Inner Mongolia Reach of the Yellow River. Proceedings of 21th International Association of Hydraulic Research Symposium on ice, Dalian, China, pp.475-482.

Hammar, L., Kerr, D.J., Shen, H.T. and Liu, L., 1996. Anchor ice formation in gravel-bedded channels. Proceedings of 13th International Association of Hydraulic Research Symposium on Ice, Beijing, pp.843-850.

Hammar, L., Shen, H.T., Evers, K.-U., Kolerski, T., Yuan, Y., and Sobaczak, L., 2002. A laboratory study on freeze up ice runs in river channels. Ice in the Environments: Proceedings of 16th International Association of Hydraulic Research Symposium on ice, 3, Dunedin,pp.36-39.

Haresign, M. and Clark, S., 2011. Modeling Ice Formation on the Red River near Netley Cut. CGU HS Committee on River Ice Processes and the Environment, 16th Workshop on River Ice, Winnipeg, Manitoba, pp.147-160.

Hicks, F., Cui, W. and Ashton, G.D., 2008. Heat transfer and ice cover decay. In: Beltaos, S.(Ed.), Chapter 4, River Ice Breakup. Water Resources Publications, LLC, Highlands ranch, Co., pp.67-123.

Hirayama, K., Terada, K., Sato, M., Sasamoto, M., Yamazaki, M., 1997. Field measurements of anchor ice and frazil ice. Proceedings of the 9th CGU-HS CRIPE Workshop on the Hydraulics of Ice Covered Rivers, Fredericton, NB, Canada

Jasek, M., 2003. Ice jam release surges, ice runs, and breaking fronts: field measurements, physical descriptions, and research needs. Canadian Journal of Civil Engineering 30, pp.113-127.

Kerr, D.J., Shen, H.T., and Daly, S.F., 2002. Evolution and hydraulics of anchor ice on gravel bed Cold Regions Science and Technology, 35, pp. 101-114.

Kempema, E., Ettema, R., and McGee, B., 2008. Insights from Anchor Ice Formation in the Laramie River, Wyoming. Proceedings of 19th International Association of Hydraulic Research Symposium on ice, Vancouver, British Columbia, Canada.Vol.1, pp.113-126.

Kolosov, M.A. and Vasiljevskiy, V.V., 2004. Prospects of winter navigation in Sayano-Shushenskoye HPS reservoir. Proceedings of 17th International Association of Hydraulic Research Symposium on ice, Saint Petersburg, Russia.

Kovachis, N., Hicks, F., Zhao, L.M. and Maxwell, J., 2010. Development of an Expert System for forecasting the progression of Breakup at Hay River, NWT, Canada. Proceedings of 20th International Association of Hydraulic Research Symposium on ice, Lahti, Finland.

Lal, A.M.W. and Shen, H.T., 1991. A Mathematical Model for River Ice Processes. Journal of Hydraulic Engineering, ASCE, 177(7), 851-867, also USA Cold Regions Research and Engineering Laboratory, CRREL Report 93-4, May 1993.

Larsen, P.A., 1969. Heat losses caused by an ice cover on open channels. Journal of Boston Society of Civil Engineers, Vol.111, No.1, pp.45-67.

Li, H., Jasek, M. and Shen H.T., 2002. Numerical Simulation Of Peace River Ice Conditions. Proceedings of 16th International Association of Hydraulic Research Symposium on ice, Dunedin, New Zealand.   Vol.1, pp.134-141.

Lier, Φ.E., 2002. Modeling of ice dams in the Karasjohka River. Proceedings of 16th International Association of Hydraulic Research Symposium on ice, Dunedin, New Zealand.

Liu, C.J., Yu, H., Ma, X.M., 2000. Ice flood in Northeast region. 15th International Symposium on Ice, Gdansk, Poland. pp. 303-312.

Liu, L.W. and Shen, H.T., 2004. Dynamics of ice jam release surges. Proceedings of 17th International Association of Hydraulic Research Symposium on ice, Saint Petersburg, Russia. pp.8.

Liu, L.W. and Shen, H.T., 2006. A two-Dimensional Comprehensive River Ice Model. Proceedings of 18th International Association of Hydraulic Research Symposium on ice, Sapporo, Japan, pp. 69-76.

Lu, S., Shen, H.T., and Crissman, R.D., 1999. Numerical study of ice jam dynamics in upper Niagara river. Journal of Cold Regions Engineering, ASCE, Vol.13, No.2, pp.78-102.

Mahabir, C., Hicks, F.E. and Robinson Fayek, A., 2002. Forecasting Ice Jam Risk at Fort Mcmurray, AB, Using Fuzzy Logic. Proceedings of 16th International Association of Hydraulic Research Symposium on ice, Dunedin, New Zealand. Vol.1, pp.91-98.

Mahabir, C.F., Hicks,F.E., and Robinson Fayek, A., 2006. Neuro-fuzzy river ice breakup forecasting system. Journal of cold Regions Science and Technology 46(2), pp.100-112.

Majewski,W. and Mrozinski, L., 2010. Ice phenomena on the Lower Vistula. Proceedings of 20th International Association of Hydraulic Research Symposium on ice, Lahti, Finland.

Malenchak, J., Doering, J., Shen, H.T. and Morris, M., 2008. Numerical Simulation of Ice Conditions on the Nelson River. Proceedings of 19th International Association of Hydraulic Research Symposium on ice, Vancouver, British Columbia, Canada.Vol.1, pp.251-262.

Massie, D.D., 2001. Neural Network Fundamentals for Scientists and Engineers, ECOS 2001, International Conference on Efficiency, Cost, Optimisation, Simulation and Environmental Aspects of Energy and Process Systems, ASME, July 4-6.

Matousek, V., 1984. Types of ice run and conditions for their formation. Proceedings of IAHR Ice Symposium, pp.315-327.

Matousek, V., 1990. Thermal processes and ice formation in rivers. Papers and Studies No.180, Water Research Inst., Prague, pp.146.

Michel, B., Mascotte, N., Fonseca, F. and Rivard, G., 1982. Formation of Border Ice in the St. Anne River. Workshop on the Hydraulics of Ice Covered Rivers-CRIPE, 1982, Edmonton, AB, Canada, pp.38-61.

Miles, T., 1993. A study of border ice growth on the Bruntwood River. M.S. Thesis, Department of Civil Engineering , University of Manitoba.

Morris, M., Malenchak, J. and Groeneveld, J., 2008. Thermodynamic Modeling to Test the potential for Anchor Ice Growth in post-construction conditions on the Nelson River. Proceedings of 19th International Association of Hydraulic Research Symposium on ice, Vancouver, British Columbia, Canada.Vol.1, pp.263-272.

Moya-Gomez, V., Popescu, I., Solomatine, D., and Bociort, L., 2013. Cloud and cluster computing in uncertainty analysis of integrated flood models, J. Hydroinform., 15, pp.55-69.

Mynett, A.E., 2002. *Environmental Hydroinformatics: the way ahead.* In: Falconer (Editor), Proc. Fifth International Conference on Hydroinformatics. IWA, Cardiff, UK, pp.31-36.

Nezhikhovskiy, R.A., 1964. Coefficients of roughness of bottom surface of slush ice cover. Soviet Hydrology, Selected Papers, No.2, pp.127-148.

O'Neil, K. and Ashton, G.D., 1981. Bottom heat transfer to water bodies in winter. Special Report 81-8. Cold regions Research and Engineering Lab., U.S.Army, Hanover,N.H.

Osterkemp, T.E., 1978. Frazil ice formation : A review. Journal of Hydraulic Division, ASCE, 104(9), pp.1239-1255.

Paily, P.P., Macagno, E.O., and Kennedy, J.F., 1974. *Winter-regime surface heat loss from heated streams.* IIHR Report No.155. Iowa Institute of Hydraulic Research, Iowa City, Iowa. pp.137.

Pariset, R. and Hausser, H., 1961. Formation and evolution of ice covers on rivers. Transaction , Engrg. Inst. Canada 5(1), pp.41-49.

Price, R.K., 2006. The growth and significance of hydroinformatics, River Basin Modelling for Flood Risk Mitigation. Taylor and Francis Group plc, London, UK, pp.93-109.

Qiu, Y., 2006. A theoretical and experimental study of anchor ice. Ph.D. Thesis , Department of Civil and Environmental Engineering , University of Manitoba, Winnipeg, Manitoba, Canada.

Qu,Y.X., and Doering, J., 2007, Laboratory study of anchor ice evolution around rocks and on gravel beds: Canadian Journal of Civil Engineering, V.34, pp.46-55, doi:10.1139/L06-094.

Rao, S.Q., Yang, T.Q., Liu, J.F. and Chen, D.L., 2012. Characteristics of Ice Regime in the Upper Yellow River in the last Ten Years. Proceedings of 21st International Association of Hydraulic Research Symposium on ice, Dalian, China, pp.390-396.

Saint-Venant, B., 1871. Theroric du mouvement non permanent des eau, avec applocation aux crues des rivieres et al, Introduction des mareees dans leurs lits. C. R. Sean. Acad. Sci., 73, pp.147-154.

She, Y. and Hicks, F., 2006. Modeling ice jam release waves with consideration for ice effects. Cold Regions Science and Technology Vol.45, pp.137-147

Shen, H.T. and Chiang, L.A., 1984. Simulation of growth and decay of river ice cover. Journal of   Hydraulic Division, Am.Soc.Civ.Eng.,110(7), pp.958-971.

Shen, H.T. and Lal, A.M.W., 1986. Growth and decay of river ice covers. Proceedings of Cold Regions Hydrology Symposium, AWRA, Fairbanks, pp.583-591.

Shen, H.T., Su, J. and Liu, L., 2000. SPH Simulation of River Ice Dynamics, Journal of Computational physics, 165(2), pp.752-771.

Shen, H.T., 2006. A trip through the life of river ice - research progress and needs. Keynote lecture. Proceedings of 18th International Association of Hydraulic Research Symposium on ice, Sapporo, Japan.

Shen, H.T., Gao, L. and Kolerski, T., 2008. Dynamics of River Ice Jam Formation. Proceedings of 19th International Association of Hydraulic Research Symposium on ice, Vancouver, British Columbia, Canada.Vol.1, pp.365-374.

Shen, H.T., 2010. Mathematical modeling of river ice processes. Cold Regions Science and Technology Vol.62, pp.3-13.

Shen, H.T., Shen, H.H. and Tsai, S.M., 1990. Dynamic transport of river ice. Journal of Hydraulic Research 28(6), pp.659-671.

Shen, H.T., Chen, Y.C., Wake, A. and Crissman, R.D., 1993. Lagrangian discrete parcel simulation of river ice dynamics. International Journal of Offshore and Polar Engineering, 3(4), pp.328-332.

Shen, H.T. and Liu, L.W., 2003. Shokotsu River ice jam formation. Cold Regions Science and Technology Vol.37(1), pp.35-49.

Shen, H.T. and Van DeValk W.A., 1984. Field investigation of St. Lawrence River hanging ice dams. Proceedings of IAHR Ice Symposium, Hamburg, pp.241-249.

Shen, H.T. and Wang, D.S., 1995. Under cover transport and accumulation of frazil granules. Journal of Hydraulic Engineering, ASCE 120 (2), pp.184-194.

Shen, H.T., Wang, D.S. and Lal, A.M.W., 1995. Numerical simulation of river ice processes. Journal of Cold Regions Engineering, ASCE 9(3), pp.107-118.

Shen, H.T. and Yapa, P.D., 1985. A unified degree-day method for river ice cover thickness simulation. Canadian Journal of Civil Engineering, pp.54-62.

Stefan, J., 1889. Uber die theorien des eisbildung insbesondere uber die eisbildung in polarmure. Wien Sitzungsberichte Akademie der Wissenschaften, Series A 42, 965-983 pt.2.

Sun, Z.C. and Sui, J.Y., 1990. Research and significance of river jams, Journal of Scientific development on earth, pp.51-54.

Svensson, U., Billfalk, L. and Hammar, L., 1989. A mathematical model for border ice formation. Cold Regions Science and Technology 16, pp.179-189.

Tesaker, E., 1975. Accumulation of frazil ice in an intake reservoir. Proceedings of IAHR Symposium on Ice problems , Hanover, NH..

Terada, K., Hirayama, K., and Sasamoto, M., 1998. Field measurement of anchor and frazil ice. Proceedings of 14th International IAHR Ice Symposium.

Tsang, G. and Lau, Y., 1995. Frazil and anchor ice laboratory investigations. Proceedings of 8th Workshop on Hydraulics of Ice Covered Rivers, Kamloops.

Tuthill, A.M., Wuebben, J.L. and Gagnon J.G., 1998. ICETHK Users Manual, Version 1, Special Report 98-11, U.S. Army Cold Regions Research and Engineering Laboratory, Hanover, New Hampshire.

U.S. Army, 1990. HEC-2 Water Surface Profiles, Hydrologic Engineering Center, U.S.Army Corps of Engineers, Davis, California.

U.S. Army, 1998. HEC-RAS River Analysis System: Hydraulic Reference Manual, Hydrologic Engineering Center, U.S. Army Corps of Engineers, Davis, California, June.

Uzuner, M.S., Kennedy, J.F., 1976. Theoretical model of river ice jams. Journal of the Hydraulics Division, ASCE 102(HY9), pp.1365-1383.

Wang, C.Q., Mynett, A.E. and Yang, J., 2012. Effect Analysis of Air Temperature Variation on the Ice Regime of the Yellow River in the last 50 years. Proceedings of 21th International Association of Hydraulic Research Symposium on ice, Dalian, China, pp.381-389.

Wang, T., Yang,K. and Guo, Y., 2008. Application of Artificial Neural Networks to forecasting ice conditions of the Yellow River in the Inner Mongolia Reach. Journal of Hydrologic Engineering, ASCE 13(9), pp.811-816.

Yamazaki, M., Hirayama, K., Sakai, S., Sasmoto, M., Kiyohara, M., and Takiguichi, H., 1996. Formation of frazil and anchor ice Proceedings of the IAHR Ice Symposium, pp.488-496.

Ye, Q.S., Doering, J. and Shen, H.T., 2004. A laboratory study of frazil evolution in a counter rotating flume, Canada Journal of Civil Engineering, Vol.31, pp.899-914.

Photos

Photo 1   Slush ice run

Photo 2   Initial land ice in Baotou reach on the Yellow River

Photo 3   Upstream freeze-up of the Yellow River (thermal freeze-up)

Photo 4   Upstream packed freeze-up of the Yellow River (narrow-river ice jam)

Photo 5   Upstream packed freeze-up of the Yellow River (wide-river ice dam)

Photo 6   Initial thermal freeze-up ice cover at Baotou

Photo 7   Alluvial land ice during breakup period in Ning-Meng reach

Photo 8   Alluvial land ice during freeze-up period in Ning-Meng reach

Photo 9   Alluvial land ice during freeze-up period in Ning-Meng reach

Photo 10   Using bombs to destroy the ice blocks

Photo 11   Dike-breach at Wuhai in December 2001

Photo 12   Rural area flooded at Wuhai in December 2001

Photo 13   Dike restoration in Duguitalakuisu County in 2008

Photo 14   Ice regime measurement at Sanhuhekou hydrometric station

Photo 15   Ice regime measurement at Bayangaole hydrometric station

Photo 16   Ice regime measurement at Toudaoguai hydrometric station

Photo 17   On-site observation at the Ning-Meng reach

Appendix A

Daily air temperature linear regression equations for 1 to 10 days

Station	Lead-time	Stepwise regression equation
	24h	$y=2.51707+0.458875*x_5+0.439663*x_7$
	48h	$y=4.15476+0.739906*x_5$
	72h	$y=4.34844+0.779242*x_5$
	96h	$y=4.17706+0.775624*x_5$
	120h	$y=4.86345+0.865138*x_5$
Linhe	144h	$y=4.88866+0.864913*x_5$
	168h	$y=5.12413+0.895269*x_5$
	192h	$y=4.66195-0.567445*x_4+0.872416*x_5$
	216h	$y=4.23553+0.803412*x_5$
	240h	$y=5.13664+0.892484*x_5$
	24h	$y=2.65572+0.284846*x_2+0.157571*x_3-0.18285*x_4+0.213649*x_5+0.316547*x_7$
	48h	$y=3.86325+0.424601*x_2+0.169584*x_3-0.298338*x_4+0.285439*x_5$
	72h	$y=3.94826+0.418498*x_2+0.206139*x_3-0.274346*x_4+0.31685*x_5$
	96h	$y=3.96516+0.433604*x_2+0.287426*x_3-0.23821*x_4+0.329573*x_5$
	120h	$y=4.14842+0.429791*x_2+0.353826*x_3+0.286953*x_5+0.175011*x_6$
Baotou	144h	$y=4.13117+0.713665*x_2+0.402872*x_3-0.580642*x_4+0.172376*x_5$
	168h	$y=3.86858+0.947414*x_2+0.47352*x_3-0.8152*x_4$
	192h	$y=4.08145+0.767617*x_2+0.493102*x_3-0.719972*x_4+0.159784*x_5$
	216h	$y=3.56859+0.799243*x_2+0.679094*x_3-0.710186*x_4+0.242096*x_6$
	240h	$y=3.43448+0.896092*x_2+0.618803*x_3$

Station	Lead-time	Stepwise regression equation
Bayangaole	24h	$y=-62.7262+0.111912*x_1+0.347819x_2+0.219722*x_3+0.209004*x_4+0.454755*x_7$
	48h	$y=-76.6386+0.140242*x_1+0.195814*x_2+0.287432*x_3+0.174831*x_4+0.551991*x_5$
	72h	$y=-105.665+0.192193*x_1+0.29721*x_3+0.288936*x_4+0.722132*x_5$
	96h	$y=-109.178+0.197893*x_1+0.309685*x_3+0.347675*x_4+0.694299*x_5$
	120h	$y=-111.035+0.201804*x_1+0.322866*x_3+0.476725*x_4+0.77419*x_5$
	144h	$y=-130.592+0.236113*x_1+0.58534*x_3+0.723009*x_5$
	168h	$y=-91.5136+0.16553*x_1+0.369706*x_2+0.577652*x_3+0.414498*x_5$
	192h	$y=-86.6515+0.156414*x_1+0.340673*x_2+0.638971*x_3+0.420628*x_5$
	216h	$y=2.74722+1.22266*x_4+0.825015*x_5$
	240h	$y=1.78728+0.489238*x_2+0.650994*x_3+0.499268*x_5$
Sanhuhekou	24h	$y=1.57428+0.409227*x_2+0.172396*x_3+0.166279*x_4+0.12711*x_6+0.396887*x_7$
	48h	$y=3.31505+0.493249*x_2+0.247276*x_3+0.220639*x_5+0.149457*x_6$
	72h	$y=3.34864+0.467792*x_2+0.337221*x_3+0.186785*x_4+0.259966*x_5+0.127059*x_6$
	96h	$y=3.25594+0.551382*x_2+0.374972*x_3+0.25653*x_5$
	120h	$y=3.61644+0.490765*x_2+0.43796*x_3+0.320081*x_4+0.27416*x_5+0.170182*x_6$
	144h	$y=3.531+0.537632*x_2+0.589369*x_3+0.30969*x_5$
	168h	$y=2.94699+0.837515*x_2+0.712759*x_3-0.462916*x_4+0.1869*x_6$
	192h	$y=2.82412+0.723505*x_2+0.843868*x_3+0.329991*x_6$
	216h	$y=2.72474+0.428953*x_2+0.680676*x_3+0.336507*x_5$
	240h	$y=2.27632+0.873378*x_2+0.949108*x_3$

Station	Lead-time	Stepwise regression equation
Toudaoguai	24h	$y=-34.7596+0.1*x_1+0.259722*x_2+0.1834*x_3+0.1237*x_4+0.55798*x_7$
	48h	$y=-52.52+0.096067*x_1+0.106187*x_2+0.24103*x_3+0.569005*x_5$
	72h	$y=-50.4854+0.0925059*x_1+0.122362*x_2+0.253745*x_3+0.594513*x_5$
	96h	$y=-38.2798+0.0703641*x_1+0.173786*x_2+0.28583*x_3+0.561723*x_5$
	120h	$y=1.30595+0.354278*x_2+0.246682*x_3+0.395381*x_5+0.163476*x_6$
	144h	$y=-66.94+0.121492*x_1+0.47325*x_3+0.732233*x_5$
	168h	$y=1.32497+0.37754*x_2+0.45466*x_3+0.367508*x_5+0.245881*x_6$
	192h	$y=1.11655+0.756169*x_2+0.734525*x_3-0.436135*x_4+0.328455*x_6$
	216h	$y=0.994008+0.743491*x_2+0.824115*x_3-0.630569*x_4+0.301029*x_6$
	240h	$y=1.0549+0.442527*x_2+0.758941*x_3+0.448273*x_5$
Shizuishan	24h	$y=-36.4542+0.0665847*x_1+0.389102*x_2+0.236635*x_3+0.333841*x_4+0.451762*x_7$
	48h	$y=3.64091+0.277778*x_2+0.281728*x_3+0.328184*x_4+0.555092*x_5$
	72h	$y=3.31774+0.330372*x_2+0.335026*x_3+0.317764*x_4+0.50507*x_5$
	96h	$y=-54.4441+0.102008*x_1+0.252491*x_2+0.394434*x_3+0.362982*x_4+0.470728*x_5$
	120h	$y=3.21745+0.399321*x_2+0.271916*x_3+0.427025*x_4+0.4586*x_5$
	144h	$y=2.78624+0.426393*x_2+0.445819*x_3+0.415261*x_5$
	168h	$y=1.7209+0.72828*x_2+0.516084*x_3+0.26525*x_6$
	192h	$y=1.42201+0.711406*x_2+0.777876*x_3+0.322378*x_6$
	216h	$y=1.11508+0.750726*x_2+0.485414*x_3$
	240h	$y=2.27327+0.535585*x_2+0.899406*x_3+0.395652*x_5$

Acknowledgements

This PhD thesis is the result of a joint research programme in River Ice Hydraulics and Simulation supported by UNESCO-IHE Institute for Water Education, Rijkswaterstaat (Ministry of Infrastructure and the Environment, the Netherlands) and the Yellow River Conservancy Commission (YRCC) of the Ministry of Water Resources (MWR), China. I would like to express my sincere appreciation to all organizations and persons that directly or indirectly stimulated me to conduct this research.

Many thanks are due to the scholarship received during 2009-2014 which enabled me to travel to the Netherlands to do my PhD research in the Department of Water Science & Engineering at UNESCO-IHE. During my stay in the Netherlands, I have been able to widen my knowledge and vision by learning from professors, colleagues and friends from all over the world.

To Professor Arthur Mynett, my promotor, my appreciation and gratitude go much further than what I can write here. It was you giving me the chance to start this PhD research, and it was you supervising me systematically and tirelessly during these years. You have been sharing your vision and wisdom, trusting my capability in doing research, giving me freedom to develop new ideas. Without you, this study would not have been finished successfully. I also acknowledge the kind collaboration with Dr. Ioana Popescu, my copromotor. You kindly helped me correcting my papers, guiding me in my ideas and trusting my capability of doing research. Thank you for being my mentor.

I very much appreciated UNESCO-IHE and its staffs for helping me in many ways during my study and stay in the Netherlands. I also have to acknowledge the Hydrology Bureau of YRCC for providing the opportunity to participate in this sandwich PhD study and supporting me with data and time to work on it. My gratitude also goes to all committee members for their valuable feedback.

My pure-hearted gratitude also goes to Ms. Jolanda Boots, the Admission and Fellowship Officer of UNESCO-IHE, for her excellent arrangement for PhD fellows and kind heartedness when I faced some difficulties during my study in the Netherlands. I have been very fortunate to have many friends who always supported me and shared nice times together with me. Thank you my friends: Dr. Li Shengyang, Dr. Zhu Xuan and her husband Chen Hui, Dr. Li Hong and her husband Dr. Wang Wen, Dr. Lin Yuqing, Dr. Ye Qinghua and his wife Wan Taoping, Fu Chao, Zuo Liqin, Wan Yuanyang, Xu Zheng, Sun Wen, Chen Qiuhan, Yan Kun, Pan Quan in the Netherlands; and Dr. Liu Jifeng, Dr. Zhang Fangxiu, Qiu Shuhui, Di Yanyan, Chen Dongling, Dr. Yan Yiqi, Tao Xin, Liu Xiaowei, Yang Jian, Zhang Yong, Li Genfeng, Long Hu, Liu Longqing, Dr. Dai Dong, and more from the Hydrology Bureau of YRCC. Thanks also to all other friends who I have not mentioned but who have helped me and encouraged me.

My greatest thanks go to my family. Thank you my parents-in-law, for your endless love, your encouragement, and always believing in my fantasies. Thank you, Kehan my wife, for always supporting me and taking care of our son Kaixiang, who is such an adorable and understanding child. Thanks to all my brothers and sisters and other relatives for being proud of me and for loving me.

Chunqing Wang

Delft, 2017

About the author

Wang Chunqing was born on 5 March 1972 in the historic city Kaifeng of Henan Province, China. He obtained his bachelor's degree in Department of Atmospheric Science of Nanjing University from September 1990 to July 1994. After graduation, he worked at the Hydrology Bureau of the Yellow River Conservancy Commission (YRCC) of the Ministry of Water Resources (MWR), where he was engaged in the meteorological information processing and weather forecasting of the Yellow River basin. From October 2001 to May 2003, he took part in the Water Scarcity Training Group sponsored by the MWR and the Netherlands government, and obtained his MSc Degree in Hydroinformatics (HI) at UNESCO-IHE. In March 2009, while working for the YRCC, he started his (part time) PhD study at UNESCO-IHE focusing on the ice regime and ice flood modelling for the Ning-Meng reach of the Yellow River. He then became director of the Hydrological and Water Resources Information Centre of the Hydrology Bureau responsible for measurement network management, meteorological forecasting, water resources allocation and ice flood forecasting for the YRCC Hydrology Bureau.

Publications

Du Xuesheng, **Wang Chunqing** and Li Genfeng, *The flood control computer system of the Yellow River,* Journal of China Hydrology, Vol.4, **1998** (in Chinese).

Wang Qingzhai, Fu Desheng and **Wang Chunqing**, *Using meteorological satellite data to observed the flood period precipitation of the Yellow River basin*, Meteorology Journal of Henan, Vol.4, **1999** (in Chinese).

Zhang Kejia, **Wang Chunqing** and Zhou Kangjun, *Application of LASG-REM to make rainstorm forecast on the San-Hua reach of the Yellow River*, Meteorology Journal of Henan, Vol.4, **1999** (in Chinese).

Wang Chunqing and Peng Meixiang, *Application of weather forecast on flood prevention in the Yellow River*, International Workshop on Flood Forecasting for Tropical Regions, Kuala Lumpur Malaysia, 14-19 June **1999**.

Zhang Kejia, **Wang Chunqing** and Zhou Kangjun, *Application of LASG-REM to make rainstorm forecast on the San-Hua reach of the Yellow River*, Proceedings of the International Symposium on Floods and Droughts for the IHP in Southeast Asia and the Pacific, Nanjing China, 18-21 October **1999**.

Wang Qingzhai, Zhao Weimin and **Wang Chunqing**, *Causing flood rainstorm forecast system on the San-Hua reach in the Yellow River*, Proceedings of the International Symposium on Floods and Droughts for the IHP in Southeast Asia and the pacific, Nanjing China, 18-21 October **1999**.

Yang Tequn and **Wang Chunqing**, Meteorological factors analysis on the affecting cut-off of the lower Yellow River, Meteorology Journal of Henan, Vol.2, **2000** (in Chinese).

Zhao Weimin and **Wang Chunqing**, *Quality evaluation techniques for multi-sources rainfall data*, Advances in Water Science, Vol.12 No.3, **2001** (in Chinese).

Yang Tequn, **Wang Chunqing** and Zhang Yong, *Estimating the mean rainfall on the area of the middle Yellow River though GMS cloud data*, Meteorology Journal of Henan, Vol.2, **2002** (in Chinese).

Peng Meixiang, Xue Yujie, **Wang Chunqing** and Long Hu, *The analysis of precipitation distribution & change of Shaan-Shan sector in middle reaches of Yellow Riverin recent 50 years*, Shanxi Meteorological Quarterly, Vol.2, **2002** (in Chinese).

Wang Chunqing, Peng Meixiang, Zhang Ronggang and Jin Lina, *Analysis on the cause of formation of weather in flood season in 2003 of the Yellow River*, Yellow River, 26(1), **2004** (in Chinese).

Wang Guoan, **Wang Chunqing** and Li Baoguo, *Comparison of Yellow River floods to the other floods in China, Southeast Asia and the world*, Proceedings of the 2nd International Yellow River Forum, Zhengzhou China, 17-21 October **2005**.

Wang Qingzhai, **Wang Chunqing** and Zhao Weimin, *Rainstorm monitoring and forecasting technique of the Yellow River basin*, China Water Publishing House, ISBN 7-5084-4197-4, November **2006** (in Chinese).

Wen Liye, **Wang Chunqing**, Zhang Ronggang, Zhaolei and Zhu Lihua, *Analysis on the cause of formation of weather in autumn flood season in 2005 of the Weihe River*, Yellow River, 28(10), **2006** (in Chinese).

Wang Chunqing, Zhang Ronggang, Liu Jifeng, Peng Meixiang, *Statistical characteristics of affecting tropical cyclone of the Yellow River basin*, Yellow River, 29(10), **2007** (in Chinese).

Peng Meixiang, **Wang Chunqing**, Wen Liye and Xue Yujie, *Ice flood causing analysis and forecasting research of the Yellow River*, Meteorological Publishing House, ISBN 978-7-5029-4397-4, November **2007** (in Chinese).

Wang Chunqing, Qiu Shuhui, Zhang Fangzhu and Marjolein De Weirdt, *Research on satellite based drought monitoring in the Yellow River basin*, Proceedings of the 3rd International Yellow River Forum, Dongying China, 15-19 October **2007**.

Dai Dong, Zhao Weimin, **Wang Chunqing** and Marjolein De Weirdt, *How to monitor and verify hydro-meteorological elements on the upper and middle Yellow River by using energy balancing method*, Journal of China Hydrology, 28(6), **2008** (in Chinese).

Tao Xin, **Wang Chunqing**, Yan Yiqi, Shi Yupin and Qiu Shuhui, *Improvement on the European real time flood forecasting operational system*, Journal of Water Resources and Water Engineering, 19(6), **2008** (in Chinese).

Wang Chunqing, Zhang Yong, Yang Jinfang, *Research on the monthly precipitation and runoff forecasting for the Heihe River basin*, Proceedings of the 4th International Yellow River Forum, Zhengzhou China, 20-23 October **2009**.

Wang Chunqing, Zhang Yong, Yang Jinfang, *Land cover classification from remotely sensed imagery using computational intelligence with application to the Heihe River basin*, Proceedings of the 4th International Yellow River Forum, Zhengzhou China, 20-23 October **2009**.

Wang Chunqing, Zhao Kun, Zhang Yong and Qiu Shuhui, *Application of weather radar rainfall estimation technology in San-Hua reach of the Yellow River*, Journal of China Hydrology, 30(2), **2010** (in Chinese).

Wang Chunqing, Arthur E. Mynett and Yang Jian, *Effect Analysis of Air Temperature Variation on the Ice Regime of the Yellow River in the Last 50 Years*, Proceedings of 21st IAHR International Symposium on "Ice Research for a Sustainable Environment", Li and Lu (ed.), Dalian, China. pp.381-389, **2012**.

Wang Chunqing, Arthur E. Mynett, Zhang Yong and Zhang Fangzhu, *Relation analysis of cold wave weather and ice regime of the Yellow River basin*, Journal of China Hydrology, Vol.32(5), Beijing, China. pp.48-52, **2012** (in Chinese).

Fu Chao, Popescu I., **Wang Chunqing**, Mynett A.E. and Zhang Fangxiu, *Challenges in modelling river flow and ice regime on the Ningxia–Inner Mongolia reach of the Yellow River, China*, Hydrology Earth System Science Vol.(18), pp.1225-1237, doi:10.5194/hess-18-1225-2014, **2014**.

Liu Jifeng, **Wang Chunqing**, Zhao Na, and Zhao Le, *Characteristic of winter discharge and its responses to climate change and human activities at Toudaoguai section of the Yellow River*, Journal of Glaciology and Geocryology, Vol.36(2), Lanzhou, China. pp.424-429, doi:10.7522/j, issn.1000 0240, **2014**.

Zhao Weimin, **Wang Chunqing**, Liang Zhongmin and Liu Xiaowei, *Research on runoff forecast and drought monitoring technology in the lower Weihe River basin*, China Water Publishing House, ISBN 978-7-5170-2714-0, November **2014** (in Chinese).

Huo Shiqingm, **Wang Chunqing** and Xu Zuoshou, *Progress of the hydrometeorological regime and forecasting of the Yellow River in recent 10 years*, Yellow River, 38(10), **2016** (in Chinese).

Wang Chunqing, Wang Pingwa, Fanminhao, Chen Dongling and Liu Jifeng, *Research on air temperature forecasting and ice regime observation technology in the Ning-Meng reach of the Yellow River basin*, China Water Publishing House, ISBN 978-7-5170-5896-0, September **2017** (in Chinese).

Printed and bound by CPI Group (UK) Ltd, Croydon, CR0 4YY

22/10/2024

01777614-0001